Creativity in product innovat.

Creativity in Product Innovation describes a remarkable new technique for improving the creativity process in product design.

Certain "regularities" in product development are identifiable, objectively verifiable and consistent for almost any kind of product. These regularities are described by the authors as "Creativity Templates." This book describes the theory and implementation of these Templates, showing how they can be used to channel the ideation process and thus enable people to be more productive and focused.

Representing the culmination of years of research on the topic of creativity in marketing, the Creativity Templates approach has been recognized as a breakthrough in such journals as *Science, Journal of Marketing Research, Management Science* and *Marketing Science*. It has been successfully implemented through workshops in international companies including Philips Consumer Electronics, Ford Motor Co., Ogilvy & Mather Worldwide, Kodak, Coca-Cola and many others.

Dr Jacob Goldenberg, senior lecturer at the Hebrew University of Jerusalem School of Business Administration, is a leading expert on creative thinking and product development. He has taught courses and workshops in inventive thinking for hi-tech companies and marketing organizations in both Israel and the United States, and has worked with companies such as Scitex, Intel, Motorola, Coca-Cola, Mastercard, Ogilvy & Mather and Rapp & Collins. He received a joint PhD degree on this topic from the School of Business Administration and the Racach Institute of Physics at the Hebrew University of Jerusalem, *Summa Cum Laude*. His bachelor's degree was in aeronautical engineering and his master's degree in mechanical engineering. Jacob has published many papers in leading journals, including *Science*, the *Journal of Marketing Research, Marketing Science* and *Management Science*.

Prof. David Mazursky is an internationally acclaimed marketing expert and a prolific author. He received his PhD in marketing from the Graduate School of Business Administration at New York University. A professor at the Hebrew University of Jerusalem, School of Business Administration, he is currently the head of the PhD committee at the university's School of Business Administration, and the director of The K-Mart Center for Retailing and International Marketing. His theories and research have been widely published in leading journals, including the *Journal of Marketing Research*, the *Journal of Consumer Research, Marketing Science, Organizational Behavior and Human Decision Processes*, the *Journal of Applied Psychology*, the *Journal of Experimental Social Psychology, Management Science* and *Science*.

Creativity in product innovation

Jacob Goldenberg

and

David Mazursky

CAMBRIDGE
UNIVERSITY PRESS

PUBLISHED BY THE PRESS SYNDICATE OF THE UNIVERSITY OF CAMBRIDGE
The Pitt Building, Trumpington Street, Cambridge, United Kingdom

CAMBRIDGE UNIVERSITY PRESS
The Edinburgh Building, Cambridge CB2 2RU, UK
40 West 20th Street, New York, NY 10011-4211, USA
477 Williamstown Road, Port Melbourne, VIC 3207, Australia
Ruiz de Alarcón 13, 28014 Madrid, Spain
Dock House, The Waterfront, Cape Town 8001, South Africa

http://www.cambridge.org

First published 2002
Fifth printing 2006

Printed in the United Kingdom at the University Press, Cambridge

Typeface Minion 11/14pt   *System* QuarkXPress™   [SE]

A catalogue record for this book is available from the British Library

Library of Congress Cataloguing in Publication data

Goldenberg, Jacob.
Creativity in product innovation / Jacob Goldenberg and David Mazursky.
     p.   cm.
Includes bibliographical references and index.
ISBN 0 521 80089 7
1. Product management.   2. Marketing – Management.   I. Mazursky, David.   II. Title.
HF5415.15.G598   2001
658.5 7–dc21   2001018106

ISBN 0 521 80089 7 hardback
ISBN 0 521 00249 4 paperback

Contents

Preface   *page* xi
Acknowledgements   xiii

Introduction: characterization and illustration of Creativity Templates   1

Part I   Theoretical framework

1   Codes of Product Evolution – a Source for Ideation   13

Market-based vs. product-based information   13
Sources of information for new products   13
Critical evaluation of market-based information at the *micro level*:
    limitations of current product users   16
Critical evaluation of market-based information at the *macro level*:
    the diffusion of awareness about a new idea   17
Implications of the S-shaped curve analysis   21
A proposition, a derivative and a dilemma   22
Information inherent in the product reflects market needs   23
The value of market research reconsidered   26
References   27

2   Revisiting the View of Creativity   29

Operational definition of creativity   29
Igniting the "creative spark"   30
Typology of research streams: the creative person, process and idea   31
Balancing surprise and regularity   35

The restricted–unrestricted scope ideation dilemma 36
Creativity Templates and other structured approaches 40
The major perspectives of the Creativity Templates approach 41
References 41

3   A Critical Review of Popular Creativity-enchancement Methods 44

Brainstorming 45
Lateral thinking 53
"Six Thinking Hats" 54
Mind mapping 54
Random stimulation 54
References 55

Part II   The Creativity Templates

4   The Attribute Dependency Template 59

An antenna in the snow – a detailed illustration 59
A disadvantage turned into an advantage 62
Generalization of the Attribute Dependency Template – an innovative
    lipstick 62
The basic principle of Attribute Dependency 63
How to compete with "Domino's Pizza" – a hypothetical case 66
Making a better candle 71
Are accidents necessary for locating ideas for new products? 73
Attribute Dependency – between attributes vs. within attributes
    dependency 74
Cycles of dependencies 75
Summary 75
References 75

5   **The Forecasting Matrix**   76

Searching for Attribute Dependency   76
Classification of variables   76
The forecasting matrix   78
Forecasting matrix – analyzing baby ointment   89
Managing the ideation process issues   94
Improving scanning efficiency through heuristics   96
Summary   97
Operational prescription   98
References   98

6   **The Replacement Template**   99

What is the Replacement Template?   99
Implementation of the Replacement Template   104
Case study 1 – a chair   109
Case study 2 – a scanner   114
Case study 3 – butter patties   116
When is exclusion appropriate?   119
Case study 4 – Nike-Air® ads   119
Case study 5 – Bally shoe ads   122
Replacement vs. Attribute Dependency   122
Operational prescription   123
References   123

7   **The Displacement Template**   124

What is the Displacement Template?   124
Displacement is not unbundling   127
Implementation of the Displacement Template   128
Observations on the Displacement Template   130
Operational prescription   132
References   133

8   The Component Control Template   134

What is the Component Control Template?   134
The thought process inherent in applying the Component Control
    Template   137
Component Control without the need for a change in the product   140
Observations on the Component Control Template   142
Operational prescription   143
References   143

Part III   A closer look at Templates

9   Templates in Advertising   147

Introduction   147
The fundamental templates of quality advertisements   148
Approaching creative advertising   152
Demonstrating Templates in advertising   153
Template distribution   163
Implications of Creativity Templates on creative execution in
    advertising   164
References   166

10   Further Background to the Template Theory   168

Space   168
Characteristics   168
Links   169
Configuration   170
Operators   171
Creativity Templates as macro operators   173
References   176

Part IV   Validation of the Templates theory

11   Demarcating the Creativity Templates   179

Mapping research: toward a product-based framework for Templates
   definition   179
Can Templates explain and predict the emergence of blockbuster
   products?   180
Can training in Templates improve creativity and quality of product
   ideas?   184
How effective are the Templates?   185
How effective is the Attribute Dependency Template?   188
Conclusions   194
References   195

12   The Primacy of Templates in Success and Failure of Products   197

Introduction   197
Predicting new product success   197
Early determinants   200
Project-level determinants of new product success   202
Hypotheses regarding the predictive power of Templates and other
   early determinants   204
Study 1: predicting success of patented products   205
Study 2: the unified model   206
Why early determinants can predict success   211
Conclusions of the empirical studies   213
Appendix   214
References   216

Index   219

Preface

Creativity in Product Innovation presents the culmination of years of research on the topic of creativity in marketing. Creativity has been a hot topic for many years in self-improvement books and articles. *Creativity in Product Innovation* brings a new dimension to the academic philosophies now beginning to emerge on the subject. This new paradigm has been recognized as a breakthrough in major scientific journals (e.g., *Science, Journal of Marketing Research, Marketing Science, Management Science* and *Technological Forecasting and Social Change*).

Breaking away from traditional postures, we posit that marketers may hear the voice of the customer by listening to the voice of the product. We further propose that the product itself contains necessary and sufficient information to serve as a basis for innovation, especially in cases of mutable and inconsistent markets. Certain regularities in product development are identifiable, objectively verifiable, learnable and consistent across product classes. These regularities, which we term *Creativity Templates*, can be used to channel the ideation process and thus enable people to be more productive and focused.

Research indicates that approximately 70 percent of successful new products match one of the Creativity Templates to be described in this book. Likewise, the failure rate of products developed according to the Templates is phenomenally low: only 8 percent as compared to a general failure rate of some 60 percent for all new products. Although the deliberations focus on the field of business, especially new product and service development, and to advertising and technology, the paradigm may bear much wider implications.

The system presented in this book allows anyone to manage a "reservoir" of defined practical mental constructs that may help tackle different problems encountered in daily life. Experiments showed that individuals trained in the *Creativity Template* approach were able to generate superior new product ideas to those generated by untrained individuals or people using rival methods – as judged by experts in their fields who were blind to the existence of Templates. In addition, most of those Template-fostered ideas were not replicated in any other ideation schematic.

This book will appeal to scholars and researchers, students of business and marketing managers, brand and product managers, consultants, business executives, as well as lay people interested in creativity and innovation. The Creativity Templates approach has already been implemented through workshops by SIT Int. in international companies such as Philips Consumer Electronics, Ford Motor Co., Ogilvy and Mather Worldwide, McCann/Erickson Worldwide, Motorola, Curver/Rubbermaid, Kodak, Scitex, Intel Corp., Coca-Cola, Mastercard and Rapp and Collins.

Acknowledgements

We wish to thank the hundreds of students and workshop participants who contributed by their remarks and activities to the formulation and consolidation of our theory. We thank all those who provided us with the knowledge and experience they had gained, which were woven into this book: Shuki Berg, Sol Efroni, Avner Egozi, Danit Einan, Omri Herzug, Ronni Horowitz, Meir Karlinski, John Kearon, Don Lehman, Amnon Levav, Haim Peres, Ginadi Pilkowski, Ed Sicfaus and Craig Stephan. Special thanks are due to Prof. Sorin Solomon for his inspired substantive contribution. We also extend our special thanks to Dr Idan Yaron, who edited the book and contributed to the crystallization of the structure, to Amnon Levav and SIT Int. for training people and validating the approach proposed in this book.

Introduction: Characterization and Illustration of Creativity Templates

Creativity Templates depict discernable, measurable and learnable regularities or patterns in innovations and novelties emergence. They enable us to understand general mechanisms of past product alterations, as well as to foresee the next alteration in the series.

The *Creativity Templates* approach is a counter-intuitive view for product emergence and a novel method for ideation, yet it does not contradict any current marketing theory. It does, however, add an important perspective to the process of product innovation by drawing on the primacy of the idea itself as a driving force toward new product success. Creativity templates are derived by inferring patterns in the evolution of products, such as those that can be inferred from the following illustrations.

Gates, computers and extraterrestrial intelligence

Thomas Alva Edison – *Life* Magazine's "Number One Man of the Millennium" – was one of the outstanding geniuses in the history of technology. Listed under his name are 1,093 US patents (including the incandescent light bulb, the phonograph and the motion picture projector). Many tales are told of this colorful personality.

One legend about Edison tells that Edison's guests were always complaining that the gate to his house opened with great difficulty, and they were required to exert great force in order to open it. Jokes were rampant about the obstinate gate and the clever inventor who could not find a way to fix it. At the end of his days, Edison was finally willing to explain: the gate was connected to a water pump, and anyone who opened the gate unknowingly pumped up water to fill Edison's private swimming pool.

Can anything be learned about creativity and innovation from this amusing story? Would information about the form of Edison's brain, his modes of thinking or his lifestyle aid us to be more creative?

We have no information about the form of Edison's brain, but this, in any case, would not help us much. For instance, scientists have recently debated whether one part of Einstein's brain (the inferior parietal region) was indeed physically extraordinary – but how can such information be useful to us in search for ideation methods?

Studying Edison's modes of thinking would not be of much use either. The following are some of his own words about the creative process: "A genius is a talented person who does his homework." "Good fortune is what happens when opportunity meets with preparation." "All you need to be an inventor is a good imagination and a pile of junk." "I didn't fail ten thousand times. I successfully eliminated, ten thousand times, materials and combinations which wouldn't work." "The three things that are essential to achievement are: hard work, stick-to-it- evenness, and common sense." And finally: "Genius is one percent inspiration and ninety-nine percent perspiration." You may have been thinking of flashes of genius, thunder and lightning – this is obviously not the case here.

Edison's lifestyle may also not be a worthy source of explanation. He had only a few months of formal education. He went to work on a train at the age of twelve, and set fire to that train. He was fired from his work as a telegrapher for almost causing a collision of two trains. As an employer, he was jealous of the most talented of his employees, and took credit for other men's work. He did not pay his bills, treated his wives badly, neglected his children, and "did not ever sleep the requisite eight hours a night."

It seems, therefore, that it is not from Edison's biography that we may learn about creativity. Is there another source of information concealed in his story? Before we answer this question, let us present another example drawn from a remote context.

Compaq Computer Corporation is a Fortune Global 100 company, the second largest computer company in the world and largest global supplier of personal computers. In mid-1999, the Compaq Corporation announced a new innovation: a Notebook® computer whose battery is recharged by typing on the keyboard. The ingenuity of this invention lies in the fact that the very activity that discharges the battery serves also to recharge it. The target population for this computer is businessmen on the move. Its benefits are obvious: its weight is reduced by eliminating the need to carry a spare battery and by reducing the size of the main battery.

Can you see any difference in the basic structure of the above two illustrations (Edison's gate and the new configuration of the Compaq computer)? In both cases, the innovators harnessed an energy source from the immediate environment (Edison's guests or Compaq end-users) in order to fulfill a necessary function (water pumping or battery recharging). This similarity may point to certain regularities or patterns (or, as we call them, Creativity Templates) which are the subject of this book.

The Creativity Templates approach precludes the need to enter the brains and the thought processes of innovators such as Edison and the Compaq engineers. It discovers and traces the regularities underlying creative ideas or products. These regularities can be conceived as codes embedded in the product itself and are revealed by observing the pattern in product evolution.

Can such patterns (or *Creativity Templates*) be applied to problems other than those involving energy sources? Let us illustrate further a brilliant innovation in the category of extended cellular phone speakers installed in cars. The research and development (R&D) team of Wirefree Ltd, a company that specializes in such speakers, have noticed an interesting contradiction: The barrier to the quality of the sound is the miniature loudspeaker, which is one of the most expensive parts in this product. Normally, we expect that improvement in the quality of a product would entail higher costs and consequently a more expensive product. Relying on the pattern in idea emergence that was identified above (Edison's gate and Compaq computer) we may expect an idea that will break this vicious circle.

But first, let us generalize further the template structure underlying this type of idea: An internal component that performs a certain function (Edison himself, a battery) is drawn from the system's configuration. In order to fulfill its function an external resource (energy from the guests in the case of the gate and energy from the user in the case of the computer) is used to replace the internal component that was drawn out. This replacement is designed in such a way that the new resource bears the same function as that of the removed component. Let us examine the speaker in view of this rule: It is quite obvious that in order to eliminate the size problem the cellular phone loudspeaker has to be removed. The remaining question is whether there is a resource in the vicinity of the system (car) that can carry the function of the removed loudspeaker. A perfect candidate would probably be the loudspeaker of the stereo system in the car itself.

The invention introduced in 1999 by Wirefree was to assign the function of the loudspeaker to the stereo system, enabling major cost savings and, at the same time, a substantial improvement of the sound (see Figure I.1).

It is interesting to note that the firm re-used this template in implementing the connection between the speaker and the radio system: Instead of producing a connector for each radio system, the appliance transmits the signals (in FM) to the radio. It turns off the radio while the telephone is activated.

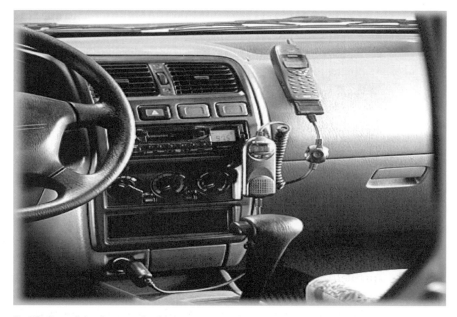

Figure I.1   The Wirefree cellular phone speaker for cars.

One abstract template surfaces from all of the above ideas, based on a code underlying them – harnessing resources from the immediate environment to replace a component that fulfills a needed function of the product. We can therefore generalize:

Replacement Template: The utilization of resources available in the immediate environment to replace a component that fulfills a needed function of the product.

This is but one code of innovation emergence that characterizes product evolution. It signifies a certain rule in evolutionary processes. Clearly, it is not the only one. Another template may be extracted from the following illustrations.

What is the shared structure of a multicolored car and pizza delivery?

Multicolored car

In 1995, Volkswagen Motor Co. launched a new model of the Polo car, named "Polo Harlequin®." The Polo, marketed until then as a solid, dependable car,

Figure I.2   The Polo Harlequin®.

acquired a new attribute: each part of the car was painted a different color (see Figure I.2). This configuration gave the car an original, mischievous look that appealed to a sizable market share. A short time after its launching, the Harlequin could be found all over Europe. The only detail that was changed in the production process was the order of assembly: instead of feeding the robot assembling the car with parts of the same color, it was fed with multicolored parts.

The curious detail about this story is that this model was originally intended as an April Fools' Day joke. Although this joke was coordinated with the PR Department of the firm who distributed multicolored posters to go with the launching (including details of two models produced in this configuration), the company did not intend to implement this concept and launch such an odd product.

To Volkswagen's great surprise, the idea captivated many customers, and a great uproar arose the next day. Orders for this (nonexistent) car started piling up in the sales department, and interest kept growing. The next step was obvious: the model was to be offered on the suddenly awakened market.

In a conventional car there is no connection between the type of outer body component of the car (e.g., doors, engine cover, top) and its color – all of the components are of the same color. The Harlequin has the characteristic of a new connection between the type of component and its color: different components have different colors. This may be illustrated in graphical form as shown in Figure I.3.

Is this procedure unique to the development of the Polo Harlequin®, or is there a recurrent pattern here that we may use in other, different and seemingly remote cases? From the realm of automobile-industry products, we shall move to an illustration from the realm of food-industry services.

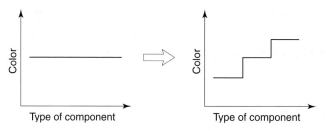

Figure I.3   Introducing a new connection in the Polo Harlequin®.

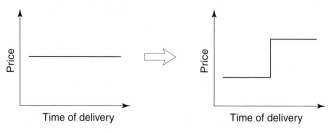

Figure I.4   Introducing a new connection in the pizza delivery.

Basically there is no difference between products and services, let us examine the same template manifestation in a famous service – pizza deliveries.

Pizza delivery

Domino's Pizza is a world leader in pizza delivery. Its success was partially due to the introduction of a novel idea: a promise to reduce the price of the pizza whenever delivery takes longer than 30 minutes. This new and original promise has caused a boom in Domino's business. In addition to the firm's obligation to fast service, there is an interesting gamble in ordering a pizza: the customer may hope that the delivery will be late and the price reduced.

Free deliveries, even a promise of fast delivery, and tasty pizzas existed before Domino's appeared on the scene. Once again an innovation can be formulated (graphically) in the same transition from a constant (straight line) to a step-function (see Figure I.4).

In this illustration, a new dependency was created by a step-function between two previously independent variables: price and time. A similar dependency could be created between two other independent variables, e.g., price and temperature. A promise would be given to deliver the pizza to the customer while hot; otherwise the price would be reduced.

Surprisingly enough, Figure I.4 presents exactly the same pattern of change as that for the Polo Harlequin®. What does this mean? Is there a common superstructure behind these ideas that may be extracted and generalized? Before drawing a conclusion, let us look at one more illustration.

The wonderful Lighthouse of Alexandria

The Lighthouse of Alexandria, built in 286–246 BC, is considered one of the Seven Wonders of the World. Many years of planning and a great engineering project led to the erection of this 134-meter tall lighthouse. It was built in order to light the seamen's way to port on stormy nights, but also to extol the name of Alexandria and its rulers. A brilliant architect directed the project. King Ptolemy II who sponsored the project wished to put a stamp of ownership on this immense asset.

The Greek architect was a genius, and put great store by the credit he would receive for his achievement. The money and vision of Ptolemy were valuable, but the architect wanted to assure recognition of his own genius and accomplishments by future generations. Such problems are solved today with the aid of a battery of lawyers dealing with intellectual or physical property, and by negotiations leading to a compromise. But those were the days of Ptolemy, when kings had no use for legal decisions in such petty matters. The architect realized that even raising the idea of credit would shorten his life considerably.

In many cases, a creative idea is expressed in the solution to a dilemma (a state in which there are two simultaneous conflicting demands). In our case the architect's dilemma was clear: the two conflicting ideas were the wish to achieve fame through the project, and the wish to continue living. The greater the fame, the shorter his life; the longer his life, the less the fame. This brilliant man needed a creative idea by which both of these requirements could be fulfilled. A compromise would be the obvious first step to take, therefore could not be termed creative even if it would solve the problem. The architect could, for example, inscribe his name in such small letters that the chance of the king seeing it would be minimized. In this case both demands would be only partially fulfilled – the credit would be small, but the risk of shortening his life would still exist (although minimized). He could also have requested in his will that his heirs should engrave his name on the lighthouse wall, but this would entail a measure of risk for them.

The architect was creative enough to find another solution. He engraved his name on the stone of the lighthouse, including a blessing for those who read his lines. He then covered the stones (and the engraving) with plaster, inscribing

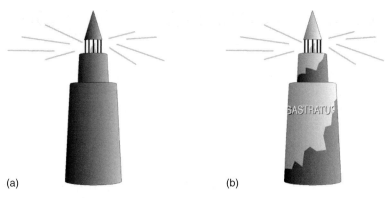

(a)   (b)

Figure I.5   The Lighthouse of Alexandria. (a) At time of construction. (b) With the passage of time.

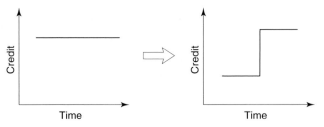

Figure I.6   Introducing a new dependency in the Lighthouse of Alexandria.

in it with great pomp the name of Ptolemy II, with praises for his deeds. Both king and architect eventually joined their forefathers, but the forces of nature worked incessantly. Erosion by sun, wind and salt air caused the peeling off and removal of the layers of plaster. The name of Ptolemy slowly disappeared, and the name of the architect – Sastratus of Cnidus – appeared in its place (See figure I.5). Thus he succeeded in bringing about his own renown as the constructor of the lighthouse (which has since collapsed as a result of an earthquake) for as long as 2,000 years after his time, without risking his life.

Legend has it that the heirs of Ptolemy enjoyed the idea so much that they did not efface the architect's name or re-plaster the area upon which he had stealthily "inherited" their ancestor's glory.

The architect of the Lighthouse of Alexandria established dependency between two formerly unrelated variables: *Credit*, nonexistent at first, was achieved only with the passing of *time* (Figure I.6). We term this pattern **Attribute Dependency Template**. The illustrations of the Polo Harlequin® (*color* and *components*) and Domino's Pizza (*price* and *time* or *price* and *temperature*) also fall in this category.

Attribute Dependency Template: Finding two independent variables and establishing dependency between them.

Another pattern used by the architect is the **Replacement Template,** which was demonstrated in the cases of Compaq portable computer and the cellular phone speaker. He utilized resources available in the immediate environment of the lighthouse (*humidity, water* and *salt*) in order to fulfill the function of *immortalizing* his name. It is consistent with the pattern obtained in the illustrations of Edison's gate (*guests' work energy* and *water pumping*), Compaq's Notebook® (*typing energy* and *recharging of battery*) and Wirefree's speakers.

The above illustrations of the Replacement Template and the systematic change between the previous and the current mode indicate that the Templates may also be used for problem-solving. In such case the goal of the problem is known a priori, and we seek ways to achieve it.

From antiquity to modernity

Templates of creative thinking taken from the past may be used for accelerating thinking about new ideas in the present.

Practically, we don't have to search for Creativity Templates in ancient Egypt or Greece. Our argument is that some templates have survived over history, and have preserved their structure even though the context has been changed. As in every evolutionary process, "the fittest survives." Adaptable historical templates – those more successful and effective – have survived well, and the information embedded in them may be used in the framework of creative thinking. Therefore, it suffices to locate templates embedded in creative ideas that have survived in the twentieth century, in which acceleration of innovative processes is witnessed and documented. The realm of new products has undergone a genuine revolution in the past 50 years, characterized by an avalanche of innovative ideas and providing evidence for the existence of Templates and possibilities for using the information embedded in them.

This introductory chapter has pointed the way to Creativity Templates and offered some clues about their meaning and operation. In Part I, the various sources used traditionally in generating new product ideas are compared with the source which underlies the Creativity Template approach. The templates approach is the one which relies on a relatively unexplored source – the information inherent in the product itself and the evolution of products over time. Part II illustrates and discusses more fully each of the major Creativity Templates. Among these templates, found to account for almost 70% of new product emergence, four templates, namely Attribute Dependency Replacement, Displacement, and Component Control, are detailed providing prescribed procedures for their implementation. Part III examines the generalizability of the notion of templates, illustrating the ways templates are derived and formulated in the context of advertising. Following identification of templates as a generalizable phenomenon, a more detailed analysis of the templates as a well-defined framework is provided. Part IV provides the empirical basis of the Creativity Templates approach – its validation in the realm of new products.

Part I

Theoretical framework

A close look at historical data reveals that new and surprising products were invented by a variety of people, some of whom had no knowledge at all about the market for the products they had developed. It appears that an alternative source for deriving new product ideas to those used at present may be to examine information embedded in the product itself. Just as market research attempts to identify trends in the marketplace on which to base a new generation of products, so market trends can be identified by analyzing **the product itself** in order to predict the basic characteristics of a new product. Because Creativity Templates inherently carry important codes for the evolution of successful new products, they can be exploited to generate a competitive advantage based on minimal market information. In Part I we describe the traditional methods of new product ideation, and assess the usefulness of market- versus product-based information, as sources of new product ideas.

Codes of Product Evolution – a Source for Ideation

Creativity is thinking up new things. Innovation is doing new things.

Theodore Levitt

Market-based vs. product-based information

The emergence of the Creativity Templates approach can be best illustrated by looking at the reality of new product innovation and the widespread methods typically used to detect and identify ideas for new products.

Introducing new products is one of the prime activities of firms and one of the most important determinants for their survival. Yet, while 25,000 products are introduced annually in the US alone, most of them are doomed to fail [1, 2]. The greatest financial loss caused by failed products occurs at the market introduction stage [3]. This state of matters may be a reason for the substantial efforts constantly devoted to "me-too" products and brand extensions [4].

Nevertheless, there is a significant correlation between innovative firms and leadership status [5]. In view of this fact, it is useful to screen at the major information sources that may determine the success or failure of new products at the ideation stage [6,7]. In marketing practice, creative ideas are highly valued and rewarded, and exploring new ideas is part of the daily activities. The high rate of failure, and the difficulties in obtaining accurate predictions of success (which are amplified when it comes to genuinely new products), are facts that need to be considered when we want to develop a marketing-oriented ideation approach.

Sources of information for new products

We tend to think that ideas arise from two sources: an "intrinsic" source based on the creative thinking by thinkers, termed "ideation," and an "extrinsic"

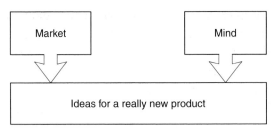

Figure 1.1   Information sources for evolving new products.

source based on market data, by searching for unfulfilled needs. Figure 1.1 describes the prevalent approach for locating original ideas for new products and services.

The marketing theory focuses on searching for ways to exhaust the extrinsic source, and presents many tools for market analysis and information extraction. However, a close look at historical data reveals some interesting facts. New and surprising products were evolved by a great variety of people, some of whom had no knowledge at all about the market for the products they had developed, and some never even imagined that the market would be interested in their inventions. Even among products that have made it into the "Hall of Fame" of innovation and creativity, most were not evolved on the basis of market research. For instance, consider the innovation of the 3M's renowned Post-It Notes® [see 8]. Despite enthusiastic employee reactions obtained in an internal company test, initial test marketing in four cities revealed reasonably acceptable results although it did not suggest outstandingly optimistic results. However, careful inspection of the data by the executives indicated a great sales potential when promotion involved free sampling and demonstration. It was that combination of selling effort that markedly raised purchase intention scores, making Post-It Notes® the most successful new product of 3M by 1984, and subsequently placing it among the top office supply products, along with copier paper, file folders and cellophane tapes [see detailed description in 9].

Likewise, Sony's Walkman® came about incidentally [see details in 10]. The earlier monophonic Pressman®, which had a recording device, was abandoned because the engineers failed to design a compact stereophonic capability. Instead of expending the necessary selling effort, it was put aside and played for the engineers' entertainment only. Only after the integration of this concept and that of light headphones was the Walkman® concept defined. Even during the first stages following the introduction, the marketers did not consider this product a potential success.

Although there are many reports of companies who have used market research methods for the improvement and adaptation of their line of products to the customer's needs, only few discuss new, breakthrough ideas arrived at solely on the basis of such research.

Over the past generation, companies have relied heavily on marketing information to guide them in new product development. The popular line of thinking asserts that new product ideas may be inferred from the customers by capturing their current needs and desires [7, 11]. By listening to the "voice of the customer," product developers expect to better forecast successful new products.

To give just one recent example, Ottum and Moore [12] declared, following a survey they conducted

although no single variable holds the key to new product performance, many of the widely recognized success factors share a common thread: The processing of market information. Understanding customer wants and needs ultimately comes down to a company's capabilities for gathering and using market information. In other words, a firm's effectiveness in market information processing – the gathering, sharing, and use of market information – plays a pivotal role in determining the success or failure of its new products.

It is true that many popular and successful products have their roots in customer feedback [for examples, see 13, 14 and 15]. This concept has also led to some very useful methods for predicting the success or failure of new product ideas and for measuring their success in terms of sales levels and market share [16].

As a result, most market researchers have focused on ways in which the market impinges on products in terms of their development, rate of innovation, market share and other performance predictors. Some of these experts have even devised methods of incorporating market considerations into the ideation and R&D stages of the product development process [13, 17, 18, 19]. [For methods of planning and designing products in response to market information see the Conjoint Analysis or the House of Quality – 20, 21.] The notion that "the market determines the fate of the product to success or failure" is also at the heart of the practice that considers consumers' preferences the primary tool in planning a new product [e.g., 22].

This line of thinking may lead us to believe that the source of ideas for new products lies solely with the customers who are responsible for the success of those products. However, while input from the market is imperative for understanding how products should be modified to meet market needs, it represents market reality at a specific time only, giving a poorer indication of *future* market needs. Nonetheless, many market research techniques continue to attempt to generate new product concepts from the elicitation and assessment of current consumer preferences and choices [e.g 13, 14].

The ability to elicit truly original concepts through market research methods has also been questioned recently. The principle arguments can be classified as (1) doubts regarding consumers' ability to provide reliable information beyond their personal experience [cf. 23] (2) the aggregate dynamics of the awareness propagation in the market dictate a low probability of an exclusive discovery of an emergent need and the subsequent introduction of an innovation to the market ("being first and alone") [for more details see 24].

Critical evaluation of market-based information at the *micro level*: limitations of current product users

Recent studies have criticized the value of market research as a source of new product ideas because of the dryness and paucity of good ideas produced [e.g., 25, 26]. These critics note that many wants lie below the surface, and that current product users are not able to express wants and needs for nonexisting products. One common method for capturing customer feedback and new product suggestions is asking buyers to describe problems with current products [7]. Yet leaps to genuinely new product ideas are unlikely to be uncovered when respondents describe problems in terms of current products.

Griffin and Hauser [21] expressed doubt about consumers' ability to foresee exactly which products the firm should develop, the details and features of the future blockbuster products and, more generally, the reliability of information provided by them about anything with which they are not personally familiar. Similarly, Griffin [23] expressed doubts about the ability of consumers to predict products worthy of development by the firm or the details and features of the future innovative products. More generally, Griffin claims that consumers cannot provide reliable information about anything with which they are unfamiliar or have no personal experience.

Von Hippel [27] summarized the problem succinctly, "Clearly, the average user of today's product has no experience with tomorrow's products." Thus, "The average user is in a poor position to provide accurate data about future products to market researchers." However, valuable information may be gathered, according to Von Hippel [see 15, 28 and 29] by being attentive to "lead users." Von Hippel indicates that these users – serving as some kind of an "ideational elite" – possess unique information about future needs. By creating solutions to their own problems, they are frequently able to predict new and successful products.

Although this approach is valuable, and well worth following, it is not qualitatively different from other, market-based solutions. In this book we are looking for information about a new product independent of the current marketplace. Altering a product according to existing market information (i.e., identifying the trend) seems to have a positive effect on the quality of an idea. However, in many cases the relevance of a trend to a product is low, partly because the idea generation process consists of an attempt to mimic other ideas rather than to generate novelty.

Beyond the assessment of the relevant studies we further posit that the collective dynamics of the market dictate a situation in which pioneering is not likely to be initiated based on market-elicited information. More specifically, we claim that the dominant factor is the *propagation of awareness* process, while the reported weaknesses of the consumer are mere representations of it, as described below.

Critical evaluation of market-based information at the *macro level*: the diffusion of awareness about a new idea

Recently, Goldenberg and Efroni [24] indicated that the propagation of awareness for an emerging need is governed by an **S**-shaped curve. The rationale for the **S** shape was first proposed by Goldenberg, Mazursky and Solomon [30]. Accordingly, consumers may become aware of a formerly unrecognized need either by self-discovery, or through communication (i.e., a consumer may be informed by another consumer who is already aware of the need). The first path can be viewed as a spontaneous change in the consumer's awareness to his or her problems and preferences (or those of other consumers), while the second is a result of an interaction with other consumers by a word-of-mouth mechanism. These two determinants are reminiscent of the variables that govern diffusion of technological innovations, echo systems, contagious diseases and sociological systems [31]. A significant difference is that the diffusion parameters are *aggregate* while the parameters that are presented here are *individual-based*.

Consider the following scenario: A new, important need emerges, unnoticed by all but a few uniquely attuned individuals who spontaneously discover the need (i.e., not as a result of any interaction with other consumers already aware of this need). Gradually, the awareness of the need is diffused by

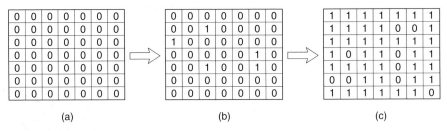

Figure 1.2 Illustration of the propagation of awareness in the Cellular Automata model – "0" represents an unaware consumer and "1" represents a consumer aware of an emerging need. The process begins in (a), the initial state, moving through (b), the early stages of awareness formation, to (c), a market with awareness to the need.

a word-of-mouth mechanism which informs unaware consumers about the new need, along with additional consumers experiencing spontaneous discovery. From a market perspective, this transition from latent to active need (characterized by market awareness) can be viewed as a propagation process, which can be modeled. For each agent, two stages of awareness can be defined: "0," representing a state of non-awareness, and "1," representing an awareness (of the emerging need). Two mechanisms govern the transition from "0" to "1:" *spontaneous discovery* – by which an agent becomes aware of the emergent need independently of other agents, and *word of mouth* – by which a proximal agent e.g., his or her neighbor, is instrumental in creating awareness. The model assumes the potential of all cells to transform from "0" to "1." As a simplifying assumption, we can also assume that the new need is a "sufficiently important new need," conforming to the intuitive understanding of a consumer as one who never "forgets" a significant need and determining an absence of depletion (irreversibility of transformation).

For each agent, the transformation probability of each mechanism may be denoted, with p as the probability for spontaneous, discovery-driven transformation and q as the probability for word-of-mouth-induced awareness. In principle, the aggregate results of p and q represent a diffusion process by which all agents ultimately undergo the transition from "0" to "1," culminating in a state of complete market awareness of the emerging need.

Figures 1.2a–c depict a two-dimensional matrix of agents (consumers) and the dynamics of the discovered propagation process at three points in time: (a) original state, (b) low awareness, and (c) almost complete market awareness of the need (saturation of awareness). This is called the Cellular Automata model.

When a real need or want is emerging, these results show that information

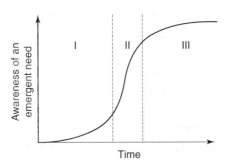

Figure 1.3   Diffusion of awareness about a new idea.

about this need propagates slowly at first, then proceeds rapidly, gradually relaxing over time. This fast propagation of awareness describes Region II of the graph. The market is finite, and therefore the rate of diffusion tapers off as the relevant public is "saturated" with the awareness of the new need or want. After some time, many people who are told about the idea will have already heard of it. Eventually, hardly anyone will be found who is not familiar with it.

Figure 1.3 describes the results of such a propagation of awareness process about an emerging need or want over time. It has three regions.

Region I represents a stage in which few people are aware of the idea for a certain product. The need for the product is latent: the potential exists, but it has yet to emerge as a recognized need. Any idea generated in this stage is likely to be perceived as creative and, once adopted, will trigger the emergence of a really new product. Hayes [32] defined creativity in terms of valuable consequences and novel or surprising outcomes. These two components of creativity are likely to be pertinent to Region I.

Region II represents the stage of fast propagation of the awareness, in which the market moves from non-awareness to saturation, and during which a conscious demand for the product is created.

Region III represents a state in which the market is saturated with information about the idea, i.e., – the consumers are aware of the need or want for the new product.

We shall limit our discussion to the extreme regions (I and III), as they are the poles of our argument:

1. Information about a product will diffuse only after the market shows an interest in it. Therefore, the product must respond to the need of as broad a population as possible.
2. The extraction of an idea from the market and its assessment by means of

a market sample (similarly to feasibility studies and studies of the predicted success of a product), are especially effective when the market is in Region III, in which the population is familiar with the idea and is able to assess it.

3. It is impossible to extract a new idea from the market when the latter is in Region I, as at this stage only a limited public knows about its existence. In this region, the statistical probability of finding consumers aware of the idea in a random sample of the population is very low.

4. In Region III, the market will provide plenty of information about the product, but this information will not be surprising, as the market is well aware of it. The competitors, able to conduct a similar market analysis, will not be surprised either. In such a situation, it is more than probable that at least one other competitor may discover the same need at the same time, as illustrated in the following discussion.

Many readers will have encountered competing products that were launched in the market almost simultaneously. The proximity in time of launching of a product by competing firms, and the business competition between them exclude the possibility that the products were copied, and we assume industrial espionage cannot account for all the cases. We believe that the manufacturers responded to a market need that was present in Region III. None of the manufacturers gained an advantage through marketing thinking; the battle rages between competing engineers, designers and planners of services to produce a better quality product at a lower price and in a shorter time, with market experts taking the position of salespeople or "information providers".

Let us portray this state of affairs by a few examples. There are firms that specialize in the promotion of new products. Inventors approach these firms with new ideas, and they provide support in the legal, operational and marketing aspects of the new product. A director of such a firm told us the following story: One morning, an inventor came to his office with a device that locks the radio-cassette in a car, a security measure against theft. The inventor had already applied for a patent on this device, and the idea was enthusiastically received in the company. Since it was simple, the device was inexpensive and the need for such a device was present. After a few days of working on the idea, a second inventor arrived, claiming that he had a revolutionary idea for the prevention of radio-cassette theft from cars. He had applied for a patent a few days after the first inventor, and his model was almost identical to the first. As the director was wondering how the two people could be so intimately coordinated his eye fell on an advertisement in the newspaper, offering a similar device for self-assembly. In this case too, a patent had been applied for.

Was this merely a rare coincidence, or is there another explanation? The answer lies in the fact that just a few weeks earlier, the insurance companies had raised the premium for insuring car radio-cassettes. A demand arose for a product that would replace the insurance costs which had risen overnight. The fact that three market-aware, imaginative people had reacted at the same time and in the same way points to the dynamics of the environment of new products as related to market needs.

When the market develops a new need this is noticed by thousands of entrepreneurs who regularly survey it for arising possibilities. When an opportunity emerges it is responded to immediately, sometimes by several companies or individual inventors concurrently. Many such examples, of simultaneous inventions responding to a new need of the market, whose inventors were not aware of each other's existence, are available. For example, Coca Cola and Pepsi Cola were developed during the same period, immediately after prohibition of alcoholic beverages was declared in the United States. These drinks contained a low dosage of codeine, had an effect similar to that of alcohol, and were marketed as an antidote to headaches [10]. Differential mathematics was developed at the same time by Leibnitz and Newton; the chip technology, micro-computers and software adaptable to different uses (such as the electronic sheet or the word processor) appeared in different parts of the world at the same time; and the incandescent light bulb and the telephone are ascribed to Edison and to Bell, although both were invented by others, who were slow or hesitant on the commercial level [33, 34].

Implications of the S-shaped curve analysis

Marketers know that they must often anticipate the competitors, and they try to capture the idea as it "climbs" the steep side of Region II. (Marketing research techniques that derive new ideas from the elicitation and assessment of existing consumer preferences and choices [e.g., 13, 14 and 15] would be suitable for Region II conditions.) The earlier they capture it, the greater the surprise effect. Yet, if the idea is captured later – as it nears the saturated region – it may be possible to save the resources needed for examining its appropriateness to the market, as in this region one may be more certain of a real need. The major differences between the two extremes – Region I and Region III – are presented in Table 1.1 below.

Let us assume that the professionals in competing firms share the same professional level. Naturally, all competitors will search for ideas in the upper part

Table 1.1 Region I versus Region III

Region I	Region III
Insufficient information in the market	Ample accessible information
Most ideas extracted from the market will be surprising	Most ideas extracted from the market will not be surprising
High probability of surprising the competition	Difficult to surprise the competition
Location of new ideas and needs through market research is ineffective	Market research maps needs and assesses them effectively

of Region II. The similarity in the function of the market experts will cause the "thinking battle" for quality, innovation and a better market position to be shifted to the development experts. Will they complete their work faster? Will the product be cheaper? Will it fulfill the need in the best way? What negative traits will it have? The development department is crucial in answering these questions, and the way they are answered will determine whether the product will succeed or fail. At this stage, the contribution of the marketing department, if it is directed by the usual line of thought described above, is poor. The function of the marketer is limited to quick and efficient transfer of information from the market to the development people – a mediation action not requiring creative thinking.

A proposition, a derivative and a dilemma

Proposition 1a

It is impossible to extract new and surprising ideas from a market in Region I, as in this stage the market is not yet aware of them and cannot provide information about them.

Proposition 1b

It is also of no use to turn to the market in Region III, as the information extracted from it will not be innovative or surprising.

Derivative

It is of no use in turning to the market for a surprising idea.

Dilemma

If market information does not provide an edge over the competitors, and precludes a surprise effect, where can we locate the market information which is indispensable for the development of ideas intended to be absorbed by the market and succeed in it?

We confine our argument to I and III, admitting the fact that in real life everything happens in II. This means that in practice our theoretical claims become underlying mechanisms that do not act alone. Yet, in reality, we see (presented in this book) that there are practically no situations in which marketing research has led a firm to introduce a truly new product without any competitiors around.

Information inherent in the product reflects market needs

Our dilemma may be resolved by way of analogy with Darwin's theory of evolution: products evolve in response to "environmental pressures" taking the form of market needs and desires. In a process resembling "the survival of the fittest", products that fail to fulfill these needs and desires disappear while products that satisfy them survive until the next change takes place. Over time, market needs and desires are "mapped" or "encoded" into a product, the configuration of which becomes a physical representation of past selection of the market or an "echo" of past customers' preferences. Moreover, since failure in need and demand satisfaction does not leave any traces, this information represents an effectively selected knowledge base. This mechanism is schematically presented in Figure 1.4. Over the years, due to the Darwinian relations between the market (environment pressure) and the product, the market-based information is captured by the products attributes.

Creativity Templates are, in fact, a well-defined sequence of operators that manipulate this knowledge base. Just as market research attempts to identify trends in the marketplace on which to base a new generation of products, so can market trends be identified by analyzing **the product itself** in order to predict the basic characteristics of a new product. Because Creativity Templates inherently carry important codes for the evolution of successful new products, they can be exploited to generate a competitive advantage based on minimal market information. They bear basic ideation schemes, and therefore may be used to help foresee new market needs while the relevant product markets are still underdeveloped.

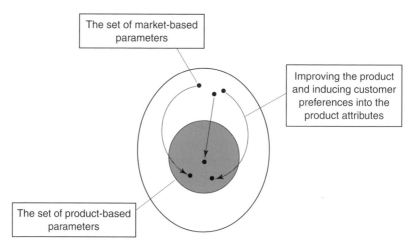

The set of market-based parameters

Improving the product and inducing customer preferences into the product attributes

The set of product-based parameters

Figure 1.4   Mapping the market-based information into the product.

In the following discussion we illustrate how market needs and desires are "encoded" into two common products: chairs and shoes.

Let us assume that we have just arrived from another world, know nothing about the market, and have never conducted market research about preferences and attitudes of people residing on this planet. However, we have the intelligence and ability to communicate well with human beings. We are surprised to discover that humans place their bodies on an object they refer to as "a chair." We observe that different chairs have legs of the same length, indicating that the height of the chair is not an arbitrary parameter, and is dependent on humans wishing to be at a certain height in relation to their environment. We may conclude that the backrest is meant to support the back and that there is a relation between the height of the chair and the height of the table next to it. The armrests of the chair support the arms, the softness of the seat is meant for comfort and is related to the body build and the preferences of the user.

We may make inferences about the needs and wants that a chair fulfills without actually asking the consumers a single question. This can be done due to the long period of evolution of the chair. Many changes and modifications have taken place over time. People have proposed ideas and improvements that the market accepted or rejected. The market itself constructs mechanisms for natural selection (in the evolutionary sense), through which a good idea survives and a poor idea disappears. A good idea is one that responds satisfactorily to an important need, leaving traces in the characteristics of the products. Over time, the product becomes a physical representation of the overarching market needs.

Table 1.2 Characteristics and benefits of a chair

Characteristics	Benefits
Length of legs	Bring body to required height
Backrest	Supports the back
Armrests	Support the arms
A chair with a central swivel leg	Ability to turn the seat sideways
Chair on wheels	Ability to move chair around the space
Adjustable backrest	Adaptability to back of sitter
Adjustable height of a swivel chair	Adaptability to height of person

Table 1.3 Characteristics and benefits of shoes

Characteristics	Benefits
Open shoe body	Aeration of foot
Heels	Increase the height of the wearer
Closed shoes	Protection from rain or dampness
Size	Adaptation of shoe to foot
Shoelaces	Fastening the shoes to the feet
Insulated shoe body	Keeping the feet warm
Orthopedic inner soles	Correct posture

Table 1.2 describes the characteristics of a chair, as well as the benefits as derived from an analysis of its characteristics.

Thus, the various market needs are engraved in the chair over time. The "negotiations" between the market and the product are mapped in the changing components of the product as they develop and improve, and reflect changes in market needs. Deciphering this code uncovers information about market needs embedded in the product.

Let us examine another product: shoes. The great variety of types of this product reflects changes that have occurred in the needs and tastes of the market. Table 1.3 shows some characteristics and benefits of shoes.

The characteristics of different shoes reflect the adaptation of the product to changes in the physical and economic environment, as well as to changing trends in the market's tastes. An examination of modifications in this product and their analysis – even based only on the above example – will provide ample information about the market's needs and tastes: the market requires shoes that adapt to changes in climate, fit the size of foot, aid posture by means of orthopedic support and will make the wearer taller. We can extract this information merely from a partial and short list of characteristics and benefits of the product.

The value of market research reconsidered

It is not always clear whether being a first mover is an advantage or a disadvantage for a specific firm's strategy [36, 35, 5]. A follower advantage may be free riding, or even gaining a "gateway for a new entry" due to technology discontinuity. However the exclusive discovery of an emerging need is a commendable achievement and a necessary step toward a possible decision to take advantage of the opportunity to control market share through pioneering status.

While recognizing the crucial role of market research in new product design and forecasting success, we argue that in some situations caution should be applied. It was shown [24] by eliciting information from consumers, that a firm operating in a market with few competitors may discover an emerging need when the gradient of the awareness propagation process is large. In that case, it is more than probable that at least one other competitor may discover the same need at the same time. Observing the saturation zone (Region III) in Figure 1.3, we conclude that despite the potential of consumer-based market research to elicit an idea which addresses a real market need, this discovery is less likely to lead to a *truly new innovative product idea* (as a greater number of competitors are simultaneously in the process of its discovery).

The above conclusion should not be interpreted to mean that market research is of little value in ideation. Market research can determine the best design of a product given market preferences, predict the product's success and ascertain the most appropriate time to launch the new product. Market researchers would be able to concentrate on these areas if we were to discover another *source* of ideas for new products, a source that would be able to predict accurately the possible future or latent needs. This source, as is obvious by now, is the product itself and its "voice."

In some instances, analysis of the current market will provide the most accessible and useful information; whereas in others, analysis of regularities observable in product changes will be more accessible and effective. As Creativity Templates inherently carry important codes for the evolution of successful new products, they can be exploited to generate a competitive advantage based on minimal market information. They bear basic ideation schemes, and may therefore be used to help foresee new market needs while the relevant product markets are still underdeveloped.

REFERENCES

1. McMath, R. M. and Forbes, T. (1998) *What Were They Thinking?* New York: Times business-Random house.
2. Bobrow, E. E. and Shafer, D. W. (1987) *Pioneering New Products: A Market Survival Guide.* New York: Dow Jones-Irwin.
3. Robertson, T. S. (1971) *Innovative Behavior and Communication* New York: Holt Rinehart and Winston Inc., p. 71.
4. Wind, J. and Mahajan V. (1997) "Issues and opportunities in new product development: an introduction to the special issue," *Journal of Marketing Research,* **34,** 1–12.
5. Cooper G. R. (1996) "New products – what separates winners from loser," in *PDMA Handbook of New Product Development.* New York: John Wiley and Sons.
6. Dolan, J. R. (1993) *Managing the New Product Development Process.* Reading, MA: Addison-Wesley.
7. Crawford, C. M. (1991) *New Products Management.* Boston, MA: Irwin.
8. Goldenberg J. and Mazursky D. (2000) "First we throw dust in the air, then we claim we can't see: navigating in the creativity storm," *Creativity and Innovation Management,* **9,** 131–143.
9. Freeman, C. and Golden, B. (1997) *Why Didn't I Think of That?* New York: John Wiley and Sons.
10. Mingo, J. (1994) *How the Cadillac Got Its Fins.* New York: HarperCollins Books.
11. Kotler, P. (1994) *Marketing Management: Analysis, Planning, Implementation, and Control.* New Jersey: Prentice Hall.
12. Ottum, B. D. and Moore, W. L. (1997) "The role of market information in new Product success/failure," *Journal of Product Innovation Management,* **14,** 258–273.
13. Lehmann, R. D., Gupta, S. and Steckel J. (1998) *Market Research and Analysis.* Homewood, IL: Richard D. Irwin.
14. Narasimhan, C. and Sen, S. K. (1983) "New product models for test market data," *Journal of Marketing,* **47,** 11–24.
15. Urban, G. L. and Von Hippel, E. (1988) "Lead user analysis for the development of new industrial products", *Management Science,* **34** (5), 569–582.
16. Lilien, G. L., Kotler P. and Moorthy, K. S. (1992) *Marketing Models.* New Jersey: Prentice Hall.
17. Freeman, C. (1982) *The Economics of Industrial Innovation.* Cambridge, MA: MIT Press.
18. Hauser, R. J. and Clausing, D. (1988) "The House of Quality," *Harvard Business Review,* May–June, 63–73.
19. Lilien, L. G. and Yoon E. (1989) "Determinants of new industrial product performance: a strategic reexamination of the empirical literature," *IEEE Transactions on Engineering Management,* **EM-36,** 3–10.
20. Urban, G. L. and Hauser, J. R. (1993) *Design and Marketing of New Products.* New Jersey: Prentice Hall.
21. Griffin, A. and Hauser, J. R. (1993) "The voice of the customer," *Marketing Science,* **12,** 1–26; 63–73.
22. Srinivasan V., Lovejoy, W. S. and Beach, D. (1997) "Integrated product design for marketability and manufacturing," *Journal of Marketing Research,* **34,** 154–163.

23. Griffin, A. (1996) "Obtaining information from consumers," in *PDMA Handbook of New Products Development*. Toronto, NY: Wieley, pp. 154–155.
24. Goldenberg J. and Efroni, S. (2001) "Using cellular automata modeling of emergence of innovations," *Technological Forecasting and Social Change* (Forthcoming).
25. Durgee, J. F., O'Connor, G. C., and Veryzer, R. W. (1996) "Using mini-concepts to identify opportunities for really new product functions," Marketing Science Institute working paper, Report No. 96–105.
26. Petroski, H. (1994) *The Evolution of Useful Things*. New York: Vintage Books.
27. Von Hippel, E. (1984) "Novel Product Concepts from Lead User: Segmenting Users by Experience," Marketing Science Institute working paper, Report No. 84–109.
28. Von Hippel, E. (1988) *The Source of Innovation*. Oxford: Oxford University Press.
29. Von Hippel, E. (1989) "New product ideas from lead users," *Research Technology Management*, **32** (3), 24–28.
30. Goldenberg, J., Mazursky, D. and Solomon, S. (1999) "Templates of original innovation: projecting original incremental innovations from intrinsic information," *Technological Forecasting and Social Change*, **11** (6), 1–12.
31. Parker M. P. (1994) "Aggregate diffusion models in marketing: a critical review," *International Journal of Forecasting*, **10**, 353–380.
32. Hayes, J. R. (1978) *Cognitive Psychology: Thinking and Creating*. Illinois: The Dorsey Press.
33. Adler, B. and Houghton, J. (1997) *America's Stupidest Business Decisions*. New York: William Morrow and Company Inc.
34. Israel, P. (1998) *Edison: A Life of Invention*. New York: John Wiley and Sons.
35. Lieberman, B. M. and Montgomery, D. B. (1988) "First Mover Advantage," *Strategic Management Journal*, **9**, 41–58.
36. Golder P. N. and Tellis, G. J. (1993) "Pioneer Advantage: Marketing Logic or Marketing Legend? *Journal of Marketing Research*, 30, (May) 158–179.

2 Revisiting The View of Creativity

Where is the life we have lost in living?
Where is the wisdom we lost in knowledge?
Where is the knowledge we lost in information?

T. S. Eliot

Operational definition of creativity

Creativity is considered the ultimate of human qualities, one of the key measures of intelligence that separates us from the rest of the animal kingdom. Our ability to create, or to innovate, is believed to be Godlike – described by some religions as one of those divine qualities endowed to man, who was created in the image of God, the Creator. Anyone who has had a spark of inspiration, a flash of genius, or even just an odd good idea, understands this seeming divinity of creative energy.

But what is creativity? And in the context of this book, what is *a creative product idea*? Numerous definitions exist for the term – more than 200 in literature alone [1, 2]. Attributes of a creative product may be: original, of value, novel, interesting, elegant, unique, surprising, endowed with power to re-order experience, not obvious, qualitatively different, etc. [e.g., 2, 3, 4, 5].

In one of our preliminary studies we distributed a questionnaire to 500 people, asking them to define the concept "a creative idea." Very few offered an exact definition, but all supplied a short list of characteristics that included adjectives such as "original," "simple," "surprising," "elegant" and "changing conventions." As the level of agreement about the list of characteristics is rather high, we may expect the people who supplied similar lists to agree on classifying an idea as creative or routine. The subjects actually did reach a large degree of agreement about such classification when presented with different ideas related to their profession.

Not only may creative products be all of that, but creative ideas are also believed to contribute to the thinking atmosphere and the enhancement of productivity, even when they do not produce immediate profit [2]. "Generating interesting designs is desirable not only because these designs will often be better than previous solutions, but also because they lead to a new way of thinking about the problem" [5].

Figuring out how to think creatively or how to "crack the code of creative thinking" has been a long sought-after goal. Research and academic studies on the topic are found across disciplines – from psychology to art to medicine.

Before we attempt to tackle the "creativity code" we must first arrive at an operational definition of creativity. Bridgman, in his manuscript *The Logic of Modern Physics* [6], indicates that one should use operational definitions in science whenever the meaning of a term in quantitative discourse is to be understood.

This book will therefore restrict the definition of a creative idea to terms of operations that can be unequivocally performed: **a creative idea is an idea about which field experts agree that it is creative.** The use of field experts to judge the creative value of ideas is indeed an accepted method in creativity research [7, 8, 9].

By virtue of this operational definition, we may ask a group of experts in relevant fields (e.g., senior marketers, advertisers or engineers) to rank different ideas according to their degree of creativity, and then examine those ideas about which there is a high level of agreement. It is our contention that if we were to amass a large quantity of ideas, let the experts grade them and choose those ranking highest in quality and creativity, we would certainly locate in them the presence of Templates of creativity.

Igniting the "creative spark"

We base our discussion on the hypothesis that creative thinking is a process one may channel, diagnose and reconstruct by use of analytical tools. This hypothesis seems surprising in view of current beliefs about creativity. As stated by Boden [10],

. . . if we take seriously the dictionary-definition of creation, "to bring into being or form out of nothing," creativity seems to be not only beyond any scientific understanding, but even impossible. It is hardly surprising, then, that some people have "explained" it in terms of divine inspiration, and many others in terms of some romantic intuition, or insight.

Indeed, many researchers hold that the creative thinking process is qualitatively different from "ordinary" day-to-day thinking [see 11, 12, 13, 14], and involves a leap which cannot be sufficiently formulated, analyzed or reconstructed. This leap is usually considered the "creative spark."

The scientific establishment refrained for many years from investigating creativity. The philosopher Carl Popper expressed the hidden assumption behind this state of mind: "Creativity is a divine spark that may not be dismantled and examined by use of scientific tools." It is not surprising, therefore, that until the 1970s no thorough scientific investigation was conducted on creativity. Hence, studying creative thinking and fostering methods that render this process more effective remained the domain of practitioners regularly handling problem-solving (engineers, advertisers, marketers, etc.), who were searching for tools for everyday use. For over thirty years many methods for fostering creativity appeared and spread, and we witnessed the growing popularity of various creativity-enhancement methods (to be discussed below). Common to all these methods was the concept that in order to ignite the creative spark all we have to do is break away from existing thought frameworks and search diligently for the non-conventional while suspending judgment and criticism.

The pioneers in breaking the "academic taboo" on research in this field were cognitive psychologists, later joined by investigators in other disciplines such as neuropsychology, artificial intelligence, engineering, education, and lately also marketing and management science. Long after the practical approval of various creativity-enhancement methods, academic research began to confirm or deny beliefs, feelings and assumptions connected with creative thinking, and to build a scientific knowledge base in this area. Among others, scholars such as Perkins [15, 16] and Weisberg [17] suggested that creativity is the outcome of ordinary thinking, only quantitatively different from everyday thinking, and does not necessarily require a qualitative leap or a creative spark. Weisberg [see 17, p. 33] summarizes the issue: "Creative thinking is not an extraordinary form of thinking. Creative thinking becomes extraordinary because of what the thinker produces, not because of the *way* in which the thinker produces it."

Typology of research streams: the creative person, process and idea

Beyond the intricate issue of definition, investigators have encountered difficulties in devising an approach through which to organize, investigate and emulate the phenomenon of creativity. Most investigators have taken one of three approaches, focusing on different aspects of creativity: the creative person, the creative process and the creative idea. All three approaches produced new assumptions and explanations that led to different strategies for "cracking the creativity code." Many of these strategies have caught the limelight and were then disproved and replaced by others.

Focusing on the creative person

This approach is designed to identify the personal profile of the creative person. Investigators who followed this lead conducted physical examinations as well as personality tests on people who were considered to be creative.

Creative people were thought to be unaccustomed to staying within set borders, and to require an open and free environment in which to think. They were expected to deviate from accepted social norms, and even to exhibit mental disturbances and mood disorders (especially manic-depressive illness or major depression). Indeed, the idea of an affinity between madness and genius is one of the oldest and most controversial of all cultural notions. Nearly 2,500 years ago Aristotle wondered, "Why is it that all men who are outstanding in philosophy, poetry, or the arts are melancholic?"

"Men have called me mad," wrote Edgar Allan Poe, "but the question is not yet settled, whether madness is or is not the loftiest intelligence – whether much that is glorious – whether all that is profound – does not spring from disease of thought – from moods of mind exalted at the expense of the general intellect." Many people have long shared Poe's suspicion that genius and insanity are entwined [18]. This suspicion was enhanced by the many examples of famous geniuses who suffered from manic-depressive illness or major depression.

Recent studies indeed indicate that a high number of established artists – far more than could be expected by chance – meet the diagnostic criteria for manic-depressive disorder or major depression. In fact, it seems that these diseases can sometimes enhance or otherwise contribute to creativity in some people.

However, it is apparent today that "most manic-depressives do not possess extraordinary imagination, and most accomplished artists do not suffer from recurring mood swings. There is very little evidence that there is a cause-and-effect relationship between manic-depressive illness and creativity. The data is mostly correlational, implying that manic-depressive illness and creativity may coexist, but not necessarily form a union. To assume, then, that such diseases usually promote artistic talent wrongly reinforces simplistic notions of the "mad genius" [18, 19].

In the early 1970s, cognitive psychologists continued to study the creative personality. They established a quantitative yardstick for measuring creativity: a creative person was someone with a great flow of ideas per unit of time [e.g., 20]. It was thought that a creative person could cull higher quality ideas from a wider set of ideas, which might be missing in a smaller set of ideas. This assumption led to a series of creativity-enhancement methods designed to boost the number of ideas suggested to solve a problem – such as brainstorming, synectics, random stimulation and lateral thinking (see Chapter 3).

More recent studies, however, have cast doubt on these methods. They show that the main difficulty faced by problem solvers is not generating a *large quantity* of ideas, but coming up with *original* ones. A large flow of ideas does not necessarily lead to original ideas. Furthermore, the very occupation with ordinary ideas seems to hamper creativity and innovative thought. No parallel between quantity and quality of ideas appears to exist [21, 22, 23].

Focusing on the creative process

This approach aimed to identify a singular, unified thought process that leads to successful creative ideas in different people. Its advocates collected written, first-person testimonies about the thought processes of creative people, and then attempted to define the general characteristics of the thought process that produces creative ideas.

The results turned up countless inconsistencies between thought processes. Every creative person reported different experiences during his or her thinking, and no tangible, defining process that characterized creative thought was apparent. The researchers concluded that free thought and lack of coercion were the foundations of creative thinking – that breaking boundaries and laws, and using free association and intuitive tools, were the portals to creative thought.

At the end of the 1980s, however, this theory also faced difficulties. Studies showed that an unrestricted scope does not necessarily lead to the golden nest of creative ideas. In fact, restricted processes of thinking are more reliable for creativity [2, 21, 24, 25, 26] (see below). On this account, many popular creativity methods that encourage large numbers of ideas have been questioned.

Focusing on the creative idea

During the 1940s, a chemical engineer named Genrich Altschuller postulated that there must be identifiable, repeated patterns or formulas underlying successful creative ideas and products. The existence of such patterns would obviate the need for searching the souls of inventors or the subjective thought process behind creative ideas [27, 28].

Altschuller's goal was to devise a systematic method to guide "ordinary" engineers toward creative solutions. By a backward analysis of more than 200,000 patents and technological inventions, he succeeded in defining more than 40 patterns of invention that he labeled "standards." Those non-intuitive patterns could be described, predicted and controlled independently of external influences. They consisted of system dynamics that could be determined solely by the intrinsic features of the products – a revolutionary idea in the field of creativity analysis.

The idea that creative solutions can be drawn from generic patterns isolated from outside influences may at first seem illogical, or even radical. The concept was wild enough to land Altschuller, a promising young engineer in the Soviet navy, in prison in 1950, under the charge of "inventor's sabotage." Before that charge, he had invented an "immobilized submarine" (a simple, reliable submarine without diving gear) and a new noxious chemical weapon that could provide a diversion in order to help a soldier trapped behind enemy lines to escape.

At a coal mine where he did forced labor during his prison term, Altschuller amazed his superiors – life-long mining experts – in resolving emergency technical situations by applying the patterns he drew from product-based information. Finally, in 1968, he was granted permission to prove his theory in a public seminar, and to meet engineers who would then become his followers. Recent studies further develop Altschuller's methods in the contexts of general problem-solving [9, 29].

The Creativity Template approach is conceptually consistent with Altschuller's attempt to uncover underlying logical patterns in the creative solutions of technological problems. The approach presented in this book extends this view of common patterns by deriving universal Templates that characterize the evolution of successful products. Moreover, while Altschuller focused his efforts on problem-solving, the Creativity Template approach focuses on new products and services in the marketing framework.

The main difference between Altschuller's approach and the Creativity Template approach is the number and parsimony of patterns or Templates proposed. Altschuller developed many. We found the number of Templates that cover the majority of successful new product innovations to be just five. Reflecting the restricted scope principle (described below), this number is much more manageable.

In the context of ideation, the Sparseness Theory [e.g., 30] holds that almost every evolutionary search for ideas is likely to yield certain common themes. The Creativity Template approach contends that a substantial part of creative behavior is guided by those abstract, fundamental schemes. Even when the creative process involves an unstructured idea-generation process, many of the ideas generated will still be definable in terms of Creativity Templates. These Templates serve as "attractors," or paths that the self-organized system tends to follow during the formation of new ideas [31].

Balancing surprise and regularity

The elements of *surprise* and *regularity* inherent in creativity intrigued many scholars. Hayes [32], for example, stressed the role of surprise in the creative process while Simon [33] and Perkins [15] stressed the importance of regularity.

Boden [34] indicated the necessary balance between *surprise* and *regularity* that corresponds to our own approach: "Unpredictability is often said to be the essence of creativity. But unpredictability is not enough. At the heart of creativity lie constraints: the very opposite of unpredictability. Constraints and unpredictability, familiarity and surprise, are somehow combined in original thinking."

Thus, the detection and utilization of templates in products does not necessarily result in undermining the surprisingness of product ideas which fit a template. Even when regularities exist creativity perception is not undermined, because it still allows ideas to be generated that most people could not or would not have arrived at [32]. Thus, products that match templates may be perceived as superior because they elicit "unrecognized familiarity;" they rely on regularities that have been proven successful in other contexts (possibly even by the same consumers) but the specific underlying regularities are not noticeable within the new context since they do not resemble the information which the other contexts contain.

Understanding of the influence of templates on creativity may be gained by considering the potential roles of templates in comparison to frameworks of knowledge transfer adopted in consumer research. Most of the relevant frameworks examine *surface similarities* between domains, rather than *structural similarities* in predicting consumer reactions. For example, growing attention has been recently devoted to analogical learning [35] which relies largely on schema-based processes as means of transferring problem solution to new problems. Analogy is accomplished through a process that generates one-to-one correspondence between elements of the representations of the base and target. If such correspondence exists, it serves as a path across which additional elements can be transferred.

When *structural similarities* are revealed, and a distinction is clearly drawn between them and surface properties, analogies from remote and unrelated domains may occur. Such inter-domain analogies may be effective when they are based on similarities embedded among the surface differences. The difficulty in easily applying a dual transfer framework in which both kinds occur concurrently, is that individuals have limited access to structural properties and they *can not distinguish between surface and structural properties* [36]. Hofstadter observed "deep" analogies, those that force the generation of new and more general rules, rather than simply providing an occasion for applying rules already learned to a new example. In much the same way, templates seem deep because the analog lacks surface similarities, which are normally observable by consumers. Templates are hidden and can be detected mainly by scientific analysis. Only when they are detected and validated externally, can they guide consumers in enhancing processing outcomes.

What this research implies is that structural and surface properties are two distinct sources of knowledge transfer, that they differ in accessibility and consumer ability to detect them, and that they function differently in the realm of knowledge transfer: surface similarities are advantageous in intra-domain transfer while structural similarities are advantageous, in far reaching inter-domain transfer.

The restricted–unrestricted scope ideation dilemma

The Templates are at the heart of another basic dilemma inherent in the ideation process. Some studies advocate that the process of generating ideas is most productive in *unrestricted* scope ideation [e.g., 37], whereas others assert that inventive thinking becomes more productive when the ideation process is confined and conducted within *restricted* scopes [15]. While the use of methods consistent with the former view is widespread [e.g., 38] scant attention has been devoted in either research or practise to methods associated with the latter view.

Because creative ideas are different from those usually arising under normal conditions, people are led to believe that in order to obtain them one has to ensure conditions dramatically different from the usual available ones. The feeling that one has to overcome mental obstacles in order to arrive at creative ideas leads to the belief that one has to ensure "total freedom" by eliminating directional guidance, constraints and restrictions [39]. The elimination of these restrictions is expected to increase the accessibility of ideas drawn and contemplated from a typically infinite space of ideas during the creativity process [e.g., 12, 40, 41].

Indeed, the most common ideation methods have traditionally tended to be informal, favoring an *unrestricted* scope ideation [e.g., 37, 42, 43]. These methods are based on the assumption that a deluge of ideas will spur further creativity [44, 45]. Another key assumption underlying many of these methods is that randomness should be stimulated and enhanced in order to facilitate a better environment for productive ideation. For example, Von Oech [46] described a set of tools devised to generate random stimuli and to allow random leaps of ideation, that are based on the premise that a random piece of information is one of the best thinking stimuli. Referring to the use of randomness-enhancing methods, Kiely [47] noted that more than one-fourth of all US companies employing over 100 people offer these employees some form of creativity training of that kind.

Many of these methods also rely on input from a creativity-generating group, rather than from individuals working alone. The ideas and suggestions produced by group interaction are supposed to suggest more ideas and accelerate the ideation process, a phenomenon called the synergetic effect or group effect.

In spite of the dominance of these approaches and related methods in the practical world, their postulated effectiveness has been eroded by many recent studies [2, 21, 22, 24, 25, 26]. Researchers have found that while wide-ranging thought increases the randomness of ideation, it decreases the usefulness of the ideas produced. Thought consciously guided by a wide set of restricted options yields more targeted and effective results.

Approaches advocating randomness have provided no research evidence that systematically assesses the efficiency of the accidental occurrence of inventions. In particular, they have not offered a way to assess the number of accidents that had no influence on new inventions, or the number of accidents that actually led to new products that failed. Creative people themselves are divided in their opinions. The composer Igor Stravinsky said, for example: "An accident is perhaps the only thing that really inspires us." On the other hand, Thomas Alva Edison observed that "None of my inventions came about totally by accident; they came by hard work."

As to informal, unrestricted thinking, one study tested the performances of problem solvers who were instructed to "break the rules, get out of the square and change paradigms," against the performances of problem solvers who were not given any instructions. The results showed no significant differences between the ideas generated by the two groups.

Moreover, it was observed that some people who are frequently asked to come up with new ideas try to find their own structured means to become more productive in ideation tasks. They may, for instance, identify patterns that are common across different contexts and apply them on an ad hoc basis within a certain product category or categories [e.g., 1, 2, 10]. Such patterns will be less transient than the random extrication of thoughts. They may also help organize the ideation process by promoting routes believed to lead to productive ideas while avoiding routes that do not. Those who adopt such cognitive strategies often have an advantage over others who treat every task on a *tabula-rasa* basis.

Dreamers

Many legends have been told about cases in which a person develops an idea as a result of a dream. One famous story appears in almost every book dealing with creativity. Friedrich August von Kekule, who discovered and defined the formula of the paraffin molecule, later tried to decipher the secret of the benzene molecule. After many years of incessant research and thinking, something happened in his life: as he was sleeping by the hearth he dreamt of balls jumping in the form of a snake swallowing its own tail. When he awoke he understood that this was the secret of benzene: contrary to what he had formerly believed, the benzene molecule is not "linear," but "circular" (hexagonal, to be exact). The lecture he delivered at the ceremony commemorating his discovery ended with the words: "Let us learn to dream, gentlemen." Koestler [12] called this incident "probably the most important dream in history since Joseph's seven fat and seven lean cows."

The idea that dreaming helps in solving problems exists alongside the belief that accidents and chance events enhance innovation. Let us examine this carefully.

When we look at new products created as a result of accidents or chance events, we may have the illusion that chance and irregular thinking do enhance creativity. But if we do that, we are ignoring the millions of unrecorded events in which accidents and chance events (or dreams) **did not** yield new products. However, we must keep in mind that most of the recorded accidents that did lead to the emergence of new products may be conceptualized by means of Creativity Templates. This could be further proof that the directed, precise and methodical use of Creativity Templates is a first-rate tool for enhancing the effectiveness of creative thinking.

As to the synergetic effect, one study asserts that it plays only a minor role in creative ideation. In a controlled experiment, ideas suggested by individuals working alone were evaluated as superior to ideas suggested in brainstorming sessions [2].

The main conclusion of such studies is that an excess of ideas and analogies obstructs the ideation process, and randomness and disorganization impede creativity. It has been realized that total freedom in problem-solving is inadequate [21, 22, 23]. "It follows," says Boden [10], "that constraints – far from being opposed to creativity – make creativity possible. To throw away all constraints would be to destroy the capacity for creative thinking. Random processes alone, if they happen to produce anything interesting at all, can result only in first-time curiosities, not radical surprises."

Originating in cognitive psychology, the *restricted* scope principle asserts that restricting the scope of an issue will enhance inventive productivity. The idea is that limiting the number of variables under consideration from infinity to a discrete, finite number will increase creativity, not decrease it. This can facilitate the ideation process and enable the individual to be more productive and focused. The capability to construct finite, discrete sub-systems that are independent of the "rest" is fundamental in natural sciences. It may also be an intrinsic feature of understanding.

Perkins [15] indicated that adherence to a cognitive frame of reference involves sensitivity to the "rules of the game." By functioning within a frame, one is better positioned to notice or recognize the unexpected. Likewise, Finke, Ward and Smith [48, p. 31] asserted that "restricting the ways in which creative cognitions are interpreted encourages creative exploration and discovery and further reduces the likelihood that the person will fall back on conventional lines of thought."

Reitman [49] observed that many problems lacking a structuring framework are *ill-defined* in that the representations of one or more of the basic components – the initial state, the operators and restrictions and the goal – are seriously incomplete, and the search space is exceedingly large. Indeed, many ill-defined problems seem difficult not because we are swamped by the task of searching through an enormous number of alternative possibilities, but because we have trouble thinking of even one idea worth pursuing. No wonder, therefore, that "total freedom" does not ensure surprise in the chosen ideas, in fact there are claims contending the contrary.

Tauber [50] pointed out that the search for new ideas typically suffers from a lack of synchronization between its two component activities – ideation, which is ill-defined, and screening, which is well-defined. Using an approach consistent with the restricted scope and Function Follows Form principles (to be discussed in detail in Chapter 5) may enhance coping with the "ill-defined" nature of ideation tasks.

Indeed, the "freedom of thought" orientation is particularly puzzling in view of attempts made by humankind throughout history to understand the regularities in nature and to utilize this knowledge for the improvement of its own well-being. Csikszentmihalyi [39, p. 5] contended that "as we ride the crest of evolution we have taken the title of creator." Yet, when it comes to planned creative thinking we choose to dissociate, unfold, untangle, randomize, etc., hoping that by the end of that task, the greatest, unsurpassed ideas will emerge.

It was therefore clear to us that certain regularities may serve as skeletons or infrastructure for screening creative ideas. By looking at these regularities a posteriori we can construct a priori skeletons for future ideas which consist of their main parameters.

During the development of the Creativity Templates approach, our decision to concentrate on the intrinsic properties of products – product-based information – was based on finding a manageable number of well-defined procedures with predictably sensible results. This would be impossible in an open, infinite system, as we would continue to encounter the "frame problem." We would be paralyzed by an infinite number of variables to consider. Or else, we would be forced to arbitrarily choose a small subset that does not necessarily behave as a system with its own rules, but is at the mercy of external perturbations not considered by the model [36].

Creativity Templates and other structured approaches

Attempts to identify relational structures have produced several frameworks conceptually analogous to the Template approach. These structures have been developed in a variety of disciplines such as linguistics [51, 52] anthropology [53], random graphics [54], venture and transitional management [33, 55] and artificial intelligence [56]. However, the background, schemes, and implications of the structures developed in these areas are essentially different.

The concept of structured creativity is embedded in a number of current ideation techniques such as Morphological Analysis [cf. 18, note also the HIT procedure, 46, 50]. Morphological Analysis is actually a group of methods that share the same structure. This method breaks down a system, product or process into its essential sub-concepts, each concept representing a dimension in a multi-dimensional matrix. Thus, every product is considered as a bundle of attributes. New ideas are found by searching the matrix for new combinations of attributes that do not yet exist. The method is used mostly in marketing, and mainly for new product ideation. The conjoint analysis is part of this group of methods [for a more detailed description see 50].

A drawback of this method relative to the Creativity Template approach is that it does not define specific guidelines for combining the parameters.

Furthermore, since the parameters are broadly defined and could lead to the generation of complex working matrices, Morphological Analysis tends to generate a large number of ideas. The absence of a prescribed reduction mechanism may compromise the selection process of the best ideas [see the discussion of limitations in 50].

A method of providing structured guidelines was recently introduced by Zaltman and Higie [57]. Their ZMET is a 10-step procedure employing personal interviews to elicit the metaphors, constructs and mental models that guide consumer thinking and behavior.

The major perspectives of the Creativity Templates approach

The Creativity Templates approach is a well-defined ideation framework integrating three major invention-enabling perspectives. The first stems from the proposition that several identified, universal Templates underlying product evolution can serve to predict new candidate products by providing the context for ideation. The second is the restricted scope principle, which channels thinking along pre-defined inventive routes in order to target creative thoughts. The third perspective is the Function Follows Form principle, which is manifested in the sequence of first proposing new configurations for a product, and then inferring the benefits, aesthetic values and other market parameters in order to create a new product idea.

Applying the approach in the application fields of this book is particularly appealing: marketing, new products and advertising are three areas in which creativity has a measurable value. Originality, surprise and innovation are part of the mix produced by every organization working in these areas.

REFERENCES

1. Dasgupta, S. (1994) *Creativity in Invention and Design: Computational and Cognitive Explorations of Technological Originality*. Cambridge: Cambridge University Press.
2. Weisberg, R. W. (1992) *Creativity Beyond The Myth Of Genius*. New York: W. H. Freeman Company.
3. Cagan, J. and Agogino, A. M. (1987) "1st PRINCE: Innovative Design of Mechanical Structures from First Principles," *AIEDAM Journal*, 1(3), 169–189.
4. Lenat, D. B. (1978) "The ubiquity of discovery," *Artificial Intelligence*, 9, 257–285.
5. Ulrich, K. T. (1988) "Computation and pre-parametric design," MIT Artificial Intelligence Laboratory, Report No. AI-TR 1043.

6. Bridgman, P. W. (1927) *The Logic of Modern Physics.* New York: The Macmillan Co.

7. Finke, R. A., Ward, T. B., and Smith, S. M. (1995) *The Creative Cognition Approach.* Cambridge, MA: MIT Press.

8. Lobert, B. M. and Dologite, D. G. (1994) "Measuring creativity of information systems ideas: an exploratory investigation," *Proceedings of the Twenty-Seventh Annual Hawaii International Conference on Systems Sciences*, pp. 392–402.

9. Maimon, O. and Horowitz, R. (1999) "Sufficient condition for inventive ideas in engineering" *IEEE Transactions, Man and Cybernetics*, **29**(3), 349–61.

10. Boden, M. A. (1991) *The Creative Mind: Myths and Mechanisms.* New York: Basic Books.

11. Guilford J. P. (1950) "Creativity," *American Psychologist*, **5**, 444–454.

12. Koestler, A. (1964) *The Act of Creation.* Arkana, UK: Penguin.

13. MacKinnon D. W. (1970) "Icreativity: a multi-faceted phenomenon," in *Creativity*, J. Roslansky (ed.) Amsterdam: North-Holland pp. 19–32.

14. Wallas, G. (1926) *The Art of Thought.* New York: Harcourt Brace.

15. Perkins, D. N. (1981) *The Mind's Best Work.* Cambridge, MA: Harvard University Press.

16. Perkins, D. N. (1988) "The possibility of invention," in *The Nature of Creativity*, R. J. Sternberg (ed.) Cambridge: Cambridge University Press pp. 362–385.

17. Weisberg, R. W. (1986) *Creativity: Genius and Other Myths.* New York: Freeman.

18. Redfield, J. K. (1995) Manic-depressive illness and creativity, *Scientific American*, **272**, 62.

19. Redfield, J. K. (1996) *Touched With Fire: Manic-Depressive Illness and the Artistic Temperament.* New York: Free Press.

20. Marschak, T., Glennan T. K., and Summers R. (1967) *Strategy for R&D Studies in Microeconomics of Development.* New York: Springer-Verlag.

21. Connolly T., Routhieaux R. L., and Schneider, S. K. (1993) "On the effectiveness of groups brainstorming: test of one underling cognitive mechanism," *Small Group Research*, **24** (4), 490–503.

22. Paulus B. P., et al. (1993) "Perception of performance in group brainstorming: the illusion of group productivity," *Personality and Social Psychology Bulletin*, **19** (1), 78–89.

23. Stroebe, W., Diehl, M., and Abakoumkin, G. (1992) "The illusion of group productivity," *Personality and Social Psychology Bulletin*, **18** (5), 643–650.

24. Bouchard T. J. and Hare, M. (1970) "Size, performance and potential in brainstorming groups," *Journal of Applied Psychology*, **54**, 51–55.

25. Diehl, M. and Stroebe, W. (1987) "Productivity loss in brainstorming groups: toward the solution of the riddle," *Journal of Personality and Social Psychology*, **53**, 497–509.

26. Diehl, M. and Stroebe, W. (1991) "Productivity loss in idea-generation groups: tracking down the blocking effects," *Journal of Personality and Social Psychology*, **61**, 392–403.

27. Altschuller, G. S. (1985) *Creativity as an Exact Science.* New York: Gordon and Breach.

28. Altschuller, G. S. (1986) *To Find an Idea: Introduction to the Theory of Solving Problems of Inventions.* Novosibirsk, USSR: Nauka.

29. Maimon O. and Horowitz R. (1997) "Creative design methodology and SIT method," *Proceedings of DETC'97: 1997 ASME Design Engineering Technical Conference* September 14–17 1997, Sacramento, California.

30. Minsky, M. (1985) *Extraterrestrials: Science and Alien Intelligence* (Edward Regis, ed.). Cambridge: Cambridge University Press.

31. Kelso, J. A. S. (1995) *Dynamic Patterns: The Self-Organization of Brain and Behavior.* Cambridge MA: MIT Press.

32. Hayes, J. R. (1978) *Cognitive Psychology: Thinking and Creating*. Illinois: The Dorsey Press.
33. Simon, H. A. (1966) "Scientific discovery and the psychology of problem solving," in *Mind and Cosmos: Essays in Contemporary Science and Philosophy*, Vol. 3. R. G. Colodny (ed.) Pittsburgh: University of Pittsburgh Press.
34. Boden, M. A. (1993) *The Creative Mind*. London: Abacus.
35. Gregan-Paxton, J. and Roedder John, D. (1997) "Consumer learning by analogy: a model of internal knowledge transfer," *Journal of Consumer Research*, **24**, 266–284.
36. Hofstadter, D. R. (1985) *Metamagical Themas*. New York: Penguin Books.
37. De Bono, E. (1992) *Serious Creativity*. New York: HarperCollins Publishers.
38. Rickards, T. (1998) "Assesing organisational creativity: an innovation benchmarking approach," *International Journal of Innovation Management*, **2** (3) 367–382.
39. Csikszentmihalyi, M. (1996) *Creativity, Flow and The Psychology of Discovery and Invention*. New York: Harper Perennial.
40. Grossman, R. S., Rodgers, B. E. and Moore, B. R. (1988) *Innovation Inc.: Unblocking Creativity in the Workplace*. Palno, TX: Wordware Publishing.
41. Parnes, S. (1992) *Sourcebook for Creative Problem Solving*. New York: Creative Education Foundation Press.
42. Klein, A. R. (1990) "Organizational barriers to creativity . . . and how to knock them down," *The Journal of Consumer Marketing*, **7** (Winter), 65–66.
43. Woodman, R. W., Sawyer, J. E. and Griffin, R. W. (1993) "Towards a theory of organizational creativity," *Academy of Management Review*, **18** (2), 293–321.
44. Batra, R., Aaker, D. A. and Myers, J. G. (1996) *Advertising Management*. Englewood Cliffs, NJ: Prentice Hall.
45. Winston, F. (1990) "The management of creativity," *International Journal of Advertising* **9**, 1–11.
46. Von Oech, R. (1983) *A Whack on the Side of the Head: How to Unlock Your Mind for Innovation*. New York: Warner Books.
47. Kiely, T. (1993) "The idea makers," *Technology Review*, (January), 33–40.
48. Finke, R. A., Ward, T. B. and Smith, S. M. (1992) *Creative Cognition*. Cambridge, MA: MIT Press.
49. Reitman, W. (1964) "Heuristic decision procedures, open constraints, and the structure of ill-defined problems," in *Human Judgments and Optimality*, W. Shelley and G. L Bryan (eds.) New York: Wiley.
50. Tauber, E. M. (1972) "HIT: Heuristic Ideation Technique – a systematic procedure for new product search," *Journal of Marketing*, **36**, 58–61.
51. Chomsky, N. (1957) *Syntactic Structures*. S-Gravenhage: Mouton and Co.
52. Eco, U. (1986) *Semiotics and the Philosophy of Language*. Bloomington, IN: Indiana University Press.
53. Levi-Strauss, C. (1974) *Structural Anthropology*. New York: Basic Books.
54. Palmer, E. M. (1985) *Graphical Evolution: An Introduction to Random Graphics*. New York: Wiley and Sons.
55. Kauffman, S. (1995) *At Home in the Universe*. Oxford: Oxford University Press.
56. Minsky, M. (1988) *The Society of Man*. New York: Simon and Schuster.
57. Zaltman, G. and Higie, R. A. (1993) "Seeing the voice of the customer: the Zaltman metaphor elicitation technique," Management Science Institute working paper, Report No. 93–114.

3 A Critical Review of Popular Creativity-enhancement Methods

In order to understand how our Creativity Templates approach compares with other creativity-enhancing methods we look at three possible approaches.

1. Methods for "management of the ideation session" (brainstorming, the Six Thinking Hats, electronic brainstorming (EBS), decision-making management, etc.) vs. methods which affect or manage cognitive processes (thinking). The Creativity Template proposed in this book approach belongs in the second category.

2. Methods trying to enhance randomness, on the assumption that it is instrumental to creativity vs. methods which are analytical and focused. Our approach belongs to the second category.

3. Generalized vs. specific methods. It is often argued that the more generalized an approach, the weaker it is; the more specific the method, the stronger it is. One may notice that the approaches in the Creativity Template category act in a narrower content world and not in general problem-solving, and should therefore be stronger. The suggested method for new products is different from that for advertising, and we claim that it has to be applied verbatim to general organizational problem-solving. Indeed, in marketing our method had to be specifically adapted to each focused content world. On the one hand, the method thus lacks generality; on the other hand, its effectiveness is increased. The lack of generality requires adaptation of the method to each content world. However, we claim that this is a worthwhile undertaking.

In the following sections we critically review some of the popular creativity-enhancement methods. The most detailed discussion will be devoted to brainstorming – probably the most widely recognized and implemented method in almost every occupation, from engineering to teaching to advertising.

Brainstorming

The knowledge underlying the concept of "brainstorming," the relative ease of its operation and assimilation in organizational contexts and the social benefits incurred in brainstorming encounters have created a ubiquity of brainstorming group sessions for problem-solving discussions in large organizations. Advertising agencies staff meet for brainstorming sessions to develop creative concepts or new advertising strategies. Engineers meet to find a solution to problems that "arrest R&D progress" and even chief executive officers (CEOs) initiate encounters with managers from various levels to review and identify new ideas for the advancement of the organization and its components. These sessions are either planned or improvised and, although sometimes guided by a professional facilitator or by one of the group participants, usually the sessions are unguided. The large variety in the nature and management of these discussions may contribute to the high acclaim received by the genre of brainstorming sessions in popularity polls relating to methods of group thinking.

The originator of brainstorming is hardly uncontested. At the Disney Studios brainstorming was an accepted method to inspire professional creativity by encouraging interactions and teamwork. In 1957, Osborne conceptualized the approach, setting down guidelines to create mental storming. Underlying Osborne's approach were a number of assumptions:

1. *People are naturally creative* Unfortunately, the ties which bind us to our routines and demanding pace of life inhibit us from flourishing creatively and generating innovative ideas. Confronted with analogies, we are released from our bonds – exposure to an analogy "disrupts" our routine reasoning and opens our mind to associations which lead to the production of original thought.

2. *Synergies* A group of people thinking together is superior to a single person thinking on his or her own. Osborne advocated that "individuals operating in a brainstorming group suggest twice as many ideas as individuals working on their own."

3. *Deferred judgment* If we eliminate the requirement to pass immediate judgment as ideas are spoken, we can gradually accumulate a pool of high-quality and original ideas which are subsequently filtered. Related to this assumption is the adage "no line of inquiry should be ruled out."

4. *Quantity leads to quality* The more we increase the number of ideas, the greater the probability of achieving a more qualitative set of ideas after filtering. As Nobel Prize winner Jonas Pulling said "The best way to get a good idea is to get a lot of ideas."

These assumptions (although formalized after the method came into practice) are the foundation on which the brainstorming method evolved in two major stages.

1. **Conceptual Brainstorming** In this first stage, a group of individuals advance ideas in no particular order, and no criteria for judgment of their merits is applied. Every idea is deemed good, and the more diverse the ideas the better. The expression of even wild and seemingly illogical ideas is invited. When participants listen to the ideas of their fellow group members, they are stimulated to create new ideas in new directions. Osborne advocated that fragments of ideas or thoughts are also welcome, as they may produce a good idea in the mind of another participant. Beyond managing the discussion, the aim of the facilitator during this stage is to create a pleasant atmosphere of deferred criticism and to encourage diversity of thought. As Doug Floyd noted, "You don't get harmony when everybody sings the same note."

Concurrently with bouncing ideas to each other, group members also respond to the ideas and suggestions of their fellow members, thus refining ideas, building new perceptions based on other ideas, or connecting a number of ideas together to consolidate a new perception.

2. **Screening** In this stage, the tens (or sometimes hundreds) of ideas are filtered to produce a reduced set which is subsequently examined and tested for economic feasibility and value. Underlying this stage are judgment and criticism, of the type deferred in the first stage. According to the classical approach to brainstorming, screening should be performed by persons other than the group members involved in the generation of the ideas. This ensures that the filtering team members are not captives of the perceptions or experiences of the first, conceptual stage. The importance of criticism at this stage is recognized, and its contribution is considered more significant after ideas have been generated during the first stage.

Brainstorming captured the hearts of organizations, industries and business concerns, many of which proudly announce their adoption of this practice, although in practice strict application of its guidelines were not universal. As brainstorming became a conventional practice and it seemed that nothing could be added to the original formula, the academic world began to examine the essential quality of the phenomenon. In the 1980s and 1990s, the method's effectiveness and relevant success factors were the subject of an increasing number of research studies. Examples of questions highlighted in this body of research are the search for the optimal number of group members and the optimal duration of the brainstorming session. The central question was, "What is brainstorming's real contribution compared to results of a *nominal group* (a group of individuals who think alone, with no contact among members)?"

One of the prominent findings was the absence of a clear advantage of the brainstorming group relative to the results of individuals who worked alone. This finding appeared repeatedly until no doubt was left in the minds of investigators: a brainstorming session does not generate more ideas or greater creativity than groups of individuals working independently [1, 2]. Is this just another case of the detachment of academic research producing surprisingly precise results that bear no relation to the reality in the field? If this was a case of a big promise that turned into a disappointing fad, we have to face the intriguing fact that this so-called management fad has struck roots and thrived for more than 30 years. Admittedly, very few management fads maintain a leading position in popularity polls for over two decades. Something is obviously going on here. The question is who is wrong?

Research indicates that most brainstorming groups did not generate more ideas than their control groups in which individuals worked alone with no contact between them. As early as 1958 (the year of Osborne's publication), the first study on this topic, by Tyler, Berry and Block empirically proved that solitary subjects produced almost twice the amount of ideas as subjects working in groups [1, 2]. Most of the numerous studies carried out over the years conclusively supported this conclusion and replicated results that stood in direct contrast to Osborne's claim. Groups were shown to be detrimental to individual productivity [see also 3].

1. The quality of the ideas themselves and their originality were inferior to the ideas generated by individuals working without any group effect.
2. The optimal number of group members for a brainstorming session was found to be 3–4, inconsistent with the conventional perception of a larger group.

These findings, which reveal that group processes have an adverse effect on creativity rather than contributing to their enrichment, created serious doubts regarding brainstorming as an effective process. Explanations suggested by researchers for the lack of success of brainstorming in laboratory experiments range from targeting elements of individual behavior in groups to the impact of brainstorming on individual problem-solving processes [see, for example, 4]. The following is a list of some of relevant factors that have been proposed:

1. *Production Blocking* In the course of idea generation, one person speaks while the others listen. As at any single moment in time only a single person can contribute ideas to the group, the scope of the potential contribution of individual group members is limited. Moreover, listening to others

express their ideas makes it harder for individuals to concentrate and develop ideas of their own. Even if they succeed in concentrating on their own ideas, recall ability is diminished.

2. *"Free riding"* As in many groups in which individual efforts are combined, brainstorming groups are not immune to attempts at free riding. In a group situation, individuals contribute their ideas to a group pool, consequently granted recognition on a group level. An opportunity is created for some members to hang onto the coattails of other, more productive members, and bask in the recognition won by group efforts without contributing personally. These free riders, who may possibly function as creative individuals in a different setting, either repeat ideas already expressed or avoid participating in the discussion.

3. *Distractions* The flow of ideas spoken aloud overwhelm individuals straining to concentrate and develop their own innovative thought. Repeated interruptions compel them to withdraw into simpler ideas that are better able to withstand the "external noise." Thus individuals are diverted to routine thoughts, mostly in an unconscious process. A related factor affecting the generation of ideas adversely in a group situation is the fact that each individual shares "thinking time" between his or her own thoughts and all the thoughts of other group members expressed aloud. Thus, group production declines relative to the sum of individual contributions in a nominal group.

4. *Deferred judgment creates a chaotic world* In a world with no judgment or criteria for assessment, individuals have no way of knowing if they are "on the right path." Rather than promoting uninhibited thought, the absence of criteria for successful ideas blocks the flow of thought. This creates two phenomena. The first is a type of helplessness and lack of direction that is typical of soldiers who have lost their bearings on a navigational task. The second is related to the cognitive loss of bearing whereby random attempts to generate ideas unfounded on prior reasoning typically produce routine, well-rehearsed thoughts. These familiar ideas are more easily produced than elusive, less developed notions.

5. *Fear of assessment* Apparently, despite instructions to the participants prior to a brainstorming session, a degree of apprehension of negative social feedback and criticism persists, inhibiting members from expressing all their original ideas.

This last proposed explanation of relatively inferior results by brainstorming seems counter-intuitive at first glance. From our experience with brainstorming, the ideas generated are so wild and preposterous that they could not be reflecting a fear of criticism. A more careful examination has shown that the reality is more complex: it seems that participants have no consternation about expressing a wild idea when there is no chance of its practicability. The rules of the brainstorming game allow far-fetched ideas, and the worst that can happen is that an idea will be received as a joke and not taken seriously as an idea for implementation. But, in the case of an original idea that is genuinely directed at the solution of the problem, criticism may be expressed (and most probably, not forgotten), as it threatens the consensus built on the solutions proposed up to that point. This creates a mechanism that filters out the most practical and dauntless of the original ideas, encouraging more banal ideas on the one hand and wilder, impractical ideas on the other.

Does all the above lead to the conclusion that brainstorming is merely a storm in a teacup? Not necessarily. Despite the laboratory experiments that invalidate the supposed effectiveness of brainstorming, field results indicate widespread adoption of the method. If we assume that such a large number of people cannot be wrong then the widespread adoption and persistence of brainstorming as an organizational practice requires an attempt to reconcile this inconsistency. Some of the explanations for the contradiction between research findings and practice may lie in what is known as the "illusion of group effectiveness."

1. **Lack of distinction between process and outcomes** The participants' ability to distinguish between their satisfaction in the process and its outcomes is distorted by the magic spell of their experience. In other words, the participants' reported satisfaction in the process is derived from their participation in the brainstorming process, leading them to over-evaluate the results compared with individuals who worked on their own and generated the same ideas. Participants in a brainstorming process indeed indicate a larger degree of satisfaction and enjoyment from the experience than nominal groups do.

2. **Group experience** Working in a group causes members to feel that something new was created in the course of the brainstorming session. In order to justify the efforts invested in the group process, individuals tend to believe that the group produced something in which all members had a part. In their evaluation of group outcomes, participating individuals do not distinguish between their ideas and the ideas of others. They are usually under the impression that they were the source of more ideas than was actually the situation. This misperception leads to the evaluation of brainstorming as an effective and satisfying process. In addition, due to the high attentiveness, group members internalize the ideas of their fellow

members, creating a false belief that more ideas were generated than would have been generated if each worked alone.

Other factors contributing to the illusion of the effectiveness of the brainstorming process cannot be identified and manipulated under laboratory conditions. A major reason for the method's popularity is found in the organizational functions served by the process, and the resulting organizational benefits that are its result [3]:

1. **Support of common organizational memory** Brainstorming sessions help organization members acquire, store, retrieve, modify and combine knowledge of various solutions to the problems they face. The sessions create opportunities to add new knowledge and solutions to the organizational memory. Furthermore, the sessions serve as an efficient means of distributing knowledge among organization members, reinforcing the knowledge of older members and imparting organizational knowledge to new members, including solutions generated previously.

2. **Diversification of ability** Participants in brainstorming frequently define it as a pleasant and fun experience. Part of the enjoyment is related to the possibility of working with others in an unrestricted manner. For most participants, the session is a social encounter, an opportunity to share experiences and discharge everyday stress. In other words, brainstorming affords participants the opportunity to experience diversity and interest that is not always present in their everyday functions.

3. **Competition over status** Brainstorming is an important organizational arena in which competition over status takes place, based on competencies of group members, who meet for a predetermined session or sessions to concentrate on a specific problem. The rules of the brainstorming session are known to all: although bad ideas are not criticized, good ideas are praised, creating an opportunity to receive less threatening feedback from organization members.

4. **Impressing clients** Brainstorming is an opportunity for an organization to convey the competencies of its members. Frequently clients are impressed by creativity expressed in meetings and they love the positive atmosphere. Organizations use this process to show their client that they understand the problem, and that they rely on a wealth of experience to arrive at the best solution. Brainstorming is also an efficient forum for the client to present problems and gather ideas from a number of participants simultaneously.

Moreover, the statement "everybody uses brainstorming" is not necessarily entirely correct. First of all, a great deal of what is called brainstorming is no more than efficient teamwork. Let us recall an example we are all familiar with:

students working together on a project. Not only are they not "brainstorming" or raising far-fetched ideas, the opposite actually occurs. They solve the problems by focusing on each problem at hand. Imagine a Harvard Business School case analysis or solving a complex mathematical function – do students raise random ideas? Do they raise ideas that are detached from each other? Do they defer judgment? Of course not. Even engineers who meet to solve problems do not conduct brainstorming sessions, though they may call it that. In practice, in most of their meetings they conduct efficient discussions, examining various alternatives and assisting each other in solving the problem. In such a process the group effect stems from focusing rather than brainstorming.

The physicist Tom Hirschfeld once said "The second attack on the same problem should be from a completely different angle." Apparently, the group encounter does have a value which is not necessarily relevant to the mechanism of brainstorming. This can be illustrated by the story of the company that decided to eliminate its coffee corner. Exhibiting great efficiency, the company appointed an individual to bring coffee and cookies to anyone who sends a request through the company's intranet. The company executives thought that by eliminating the need for the engineers to leave their desks their efficiency would be improved, and they could also convey the consideration that the company shows its employees by making it more convenient for them. After a number of months, the executives found that eliminating the random meetings at the coffee machine stopped the exchange of knowledge and opinions, which had a detrimental effect on the engineers, and their productivity fell.

This story indicates a possible means of reconciling our conflicting data. Perhaps the brainstorming method is not particularly effective for the generation of new ideas, but apparently it, or the encounter it creates, exposes previously untapped ideas of the organization's members. Imagine that you are facing a problem similar to one previously solved by a member of the same organization on the floor below. In an encounter such as brainstorming, he or she would be able to share the accumulated knowledge, the successful and failed tests that were conducted, and perhaps the directions that had been examined and found potentially beneficial.

If so, how can we reap the benefits of brainstorming while avoiding its relative shortcomings, as found in laboratory experiments? In this context, we would like to present some new insights relating to brainstorming and suggest how a more advanced version of this approach may be used to the benefit of organizations. In EBS, a recently introduced electronic version of brainstorming [see 5], brainstorming sessions – rather than taking place aloud and in a single location – are conducted by virtually merging nominal groups and opening a new channel of idea-sharing and knowledge transference. Each organization member sits at his or her own desk, electronically connected to the others (sometimes from other firms or organizations). Participants generate ideas on their own and send them to the general pool, while continuing to develop their train of thought and generate more ideas on the theme. When they feel that they are ready to investigate the ideas of others, they download them from the pool. These inputs lead to new responses on their part or to further development or generation of ideas. This virtual neuron storm has a number of advantages chiefly that individual members can control their retrieval of the ideas of other participants and choose the timing for reflection on these inputs, thus avoiding the distraction effect. Outside observers can add new instructions or guidelines. In a global brainstorming process that was conducted through the Internet at the initiation of a well-known food company, corporate executives observed the exchange of ideas in real time, on a giant screen. They concurrently conducted a discussion to assess and filter the various ideas and even made real-time decisions.

Empirical findings reflect the success of EBS, both absolutely and relative to "regular" brainstorming. Large groups of participants generated more ideas than individuals working alone (in nominal groups). In addition, the quality of the EBS-generated ideas was rated as higher. The size of the EBS group enables a quick and sharp compilation of the conceptual capital in the organization. Apparently, the absence of inhibiting social phenomena that characterize group processes during brainstorming and the ability to concentrate and develop a train of thought without interruption, avoid the pitfalls of brainstorming while enhancing its benefits. Although it is too early to claim that advanced EBS has made significant inroads into managerial practice at the cost of traditional brainstorming, its superiority has been sufficiently demonstrated to allow us here to seriously recommend its integration in the problem-solving modes of organizations. Nevertheless, we should remember that despite EBS's superiority in idea generation, it lacks the social encounter that has been found to be an important side-benefit of conventional brainstorming encounters.

Other interesting findings related to recent creativity studies (not conducted in the context of group research) relates to the perception that the constraints of a problem encourage the production of more creative ideas [see 6]. Constraints provide a focus for the brainstorming session and transform it into "focused-storming." Brainstorming synergies are not a result of distractions. They are based on a number of minds working under well-defined direction. In this context, we note that studies of group decision-making processes indicate that when a discussion is well managed according to a decision-making model, the group has a higher value than the sum of its members. Findings such as these lead us to hypothesize that when the problem is constrained and well defined, even subject to complete freedom of expression, anarchy is not a possible outcome. In our opinion, a brainstorming session should be conducted to generate solutions to well-defined problems, with a clear set of criteria for success. In this case, information exchange, conceptual capital and directing the storm of ideas to a target-focused channel contribute to organizational performance. Conversely, when problems are ambiguous and ill defined, such as problems relating to new product innovation, focused storming is preferred.

In order to make the most of brainstorming it is imperative to define the problem at hand and the goals of the encounter beforehand, and in a precise manner. Brainstorming is one of the easiest methods to implement, and although many different ways of doing this are available, these should be context-consistent and fit the situation. It is important to remember that brainstorming is not always the only or the preferred option.

Lateral thinking

This is another widely used method popularized by Edward De Bono [7]. The philosophy behind this approach may be conveyed by the analogy of digging a hole. While structural thinking is analogous to digging down in depth, lateral thinking is analogous to the search for a new spot to start digging. De Bono offers several provocative paths to force the thinker to consider different options for solution. Some of his techniques include inverting the situation, altering the situation to make it provocative, and considering interesting directions just because they are interesting (even where no benefit is seen).

Contrary to brainstorming, here the group does not take primacy; the decision about the thinking path to be followed is more important. De Bono emphasizes that every creative idea is logical a posteriori, even if it looks illusive a priori. Many large organizations have reported successful cases of the use of lateral thinking, mostly for managerial problems.

"Six Thinking Hats"

This is also a De Bono method, but it differs from lateral thinking in that it is directed at a group rather than an individual. It draws from group dynamics and is similar to brainstorming, although it proposes a rigidly structured discussion.

This method uses six different "natures" of thought, each represented by a different colored "hat" (real or imaginary). For example, a green hat represents creativity. When the leader of the discussion announces that the green hat is in use, the group participants are limited to suggesting creative ideas or views. The red hat represents emotions. Only when the red hat is in use may the participants present their personal feelings regarding a process or an idea. The secret of the successful implementation of this method is the order of the hats. This method was reported to be successful in several case studies of managerial problems [8].

Mind mapping

This method calls for a free association and flow of thoughts. An individual is instructed to draw a circle at the center of the paper and write in it a short description of the problem. He or she must then draw new circles around the page representing associations (*not* solutions) to the problem, which are linked to the first circle. Each new circle is now the origin for a new bundle of associations, which are noted again in new circles. When the paper is covered with circles (all linked to each other), the problem solver may look at the map and examine one small cluster of circles in order to try to find a solution. There are no reports of tests conducted to estimate the efficiency of this method.

Random stimulation

Sometimes presented as a complementary method to mind mapping, random stimulation posits that, if forced into the context of a problem, a remote analogy can sometimes stimulate a chain reaction of new thoughts and liberate a fixation. The method advocates choosing a random word (e.g., nail) and focusing thoughts on how that word could be part of a solution to the problem (e.g., find a new schedule of production). The method is used mainly in writing and the arts. There are no reports of tests conducted to estimate the efficiency of this method.

REFERENCES

1. Diehl, M. and Stroebe, W. (1987) "Productivity loss in brainstorming groups: toward the solution of the riddle," *Journal of Personality and Social Psychology*, **53**, 497–509.
2. Diehl, M. and Stroebe, W. (1991) "Productivity loss in idea-generation groups: tracking down the blocking effects," *Journal of Personality and Social Psychology*, **61**, 392–403.
3. Sutton, R. I. and Hargadon, A. (1996) "Brainstorming groups in context: effectiveness in a product design firm," *Administrative Science Quarterly*, **41**, 685–718.
4. Paulus, B. P., et al. (1993) "Perception of performance in group Brainstorming: the illusion of group productivity," *Personality and Social Psychology Bulletin*, **19** (1), 78–89.
5. Gallupe, R. B. et al. (1992) "Electronic Brainstorming and Group Size," *Academy of Management Journal*, **35**, 350–69.
6. Finke, R. A., Ward, T. B. and Smith, S. M. (1992) *Creative Cognition.* Cambridge, MA: MIT Press.
7. De Bono, E. (1970) *Lateral Thinking: Creativity Step by Step.* New York: Harper and Row.
8. De Bono, E. (1985) *Six Thinking Hats.* Boston: Little, Brown.

Part II

The Creativity Templates

Creativity Templates are codes embedded in the product itself and in trends observed in its evolution. Those Templates that are more successful and effective are likely to underlie products that survived well. Thus, the Templates may be used in the framework of creative thinking. The well-defined sequence of operations that underlie the change between previous and current product versions enables the construction of a prescribed procedure of invention. In Part II the derivations of four of these templates – Attribute Dependency, Replacement, Displacement and Component Control – are described and illustrated. Subsequently, for each of the four templates an operational prescription is provided.

4 The Attribute Dependency Template

This template was illustrated briefly in the introduction: the Polo Harlequin®, Domino's pizza delivery and the dynamization of the credit in the lighthouse are all examples of ideas that share the structure of this Template. In this chapter we will focus on this Template and on ways to implement it successfully in new product ideation.

An antenna in the snow – a detailed illustration

The following example shows how illusive creative ideas may often be. A company specializing in the production of transmission and reception sets participated in a bid for the production of military receiving antennas in a region where winter temperatures reach a low of −40°C. The company itself is located in a warm climate where even light snow is a rare event. It may be for this reason that the company engineers forgot to take into account a common phenomenon in the target market: The accumulation of ice on the antenna causes an overload on the pole, which may cause it to buckle down and collapse (see Figure 4.1). They therefore designed a light pole, which was not quite suited for the task. Ironically, because of the importance assigned to the weight issue (the pole should be light as it had to be carried by a team of three soldiers), the army accepted this bid. Actually, the weight of the proposed pole was half of what was needed to prevent it buckling and when it was realized that the weight of the pole had to be doubled the company had no choice but to redesign it. Now consider the nature of this situation, which is in fact identical in its logic to the case of the lighthouse: It is a design dilemma attempting to overcome two conflicting requirements. The dilemma is formulated graphically in Figure 4.1.

Could a lesson be learned from the Lighthouse of Alexandria that would be relevant to the antenna, even though there is little similarity between antenna poles and lighthouses, and the parameters of dilemma faced in each

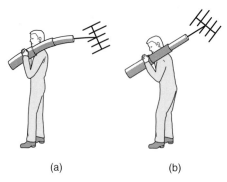

(a)   (b)

Figure 4.1   The dilemma of the antenna designer. (a) A light but weak pole. (b) A strong but heavy pole.

of them differs considerably? If indeed something can be learned, it may illustrate the practicality of Creativity Templates extracted from historical information.

As in the lighthouse case, here too, there is a contradiction between two conflicting requirements. On the one hand, the pole must be sturdy enough to withstand the weight of the ice accumulating on the antenna; on the other hand, strengthening it would make it difficult to carry around, which would breach the contract terms. There is, therefore, a basic similarity in the structure of the dilemma posed in the two cases. Is it possible that there is also a basic similarity in the solution for the two dilemmas?

As in the case of the lighthouse, two independent variables were chosen. One was time (as in the architect's choice); the other had to be clearly related to the problem – the strength or weight of the pole. Let us describe the usual case (reflecting the static approach), where there is no dependency between the time element and the strength of the pole. As with the lighthouse – where there was no need to immortalize the architect during his lifetime – there is no reason for the pole to be strong before the need arises (i.e., there is no reason to strengthen the pole before ice starts amassing on it). Thus, it is possible to add a dependency in the form of a step-function, as in the case of the lighthouse (see Figure I.6).

Here, too, we see that the "dynamization" of the pole removes the major difficulty from the logical problem. The pole will be light in structure (therefore weaker than required), and will be easily carried and moved. After installation, when a load of snow may weigh the antenna down, the pole must become stronger so as to withstand it (see Figure 4.2).

How then, can one magically render the pole stronger? It is out of the question to cover the pole with an extra layer (such as concrete) at the beginning

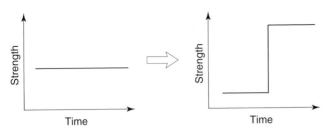

Figure 4.2   Attribute Dependency in an antenna pole.

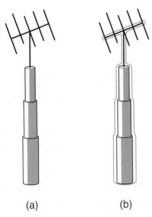

(a)   (b)

Figure 4.3   Solution of the antenna problem. (a) Light, mobile antenna. (b) Antenna strengthened by layer of ice.

of the winter. This idea is obviously both non-creative and senseless: the heavy concrete must be carried, just like the pole itself.

The Replacement Template (presented briefly in the introduction, which will be elaborated in Chapter 6) comes to our aid. According to the principle upon which the Template is based, the strengthening material must be located in the area of the pole and at the time when the load on the antenna increases. Except for air and soil, only snow and ice abound around the pole – and they are the cause of the problem. Figure 4.3 shows a situation in which the problem (ice) is also the solution.

By creating a rough surface to the pole, we can cause ice to adhere to the pole, not only to the antenna. Recall that ice is an especially strong material (a tank can easily move over a frozen lake surface with an ice layer of 50 cm); thus the ice-coated pole will be strong enough to carry the ice-loaded antenna.

Examining the examples so far may raise the question of whether dynamization is the only possibility for creating a dependency to the pole (i.e. Attribute Dependency). The introduction of a dependency to the pole is not

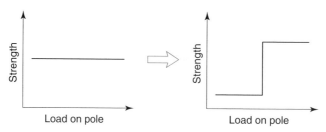

Figure 4.4   Solution scheme of antenna problem formulated by strength and load variables.

necessarily time dependent, since another dependency can be defined in this framework between the load variable on the pole and its strength variable: As the load becomes heavier, the pole becomes stronger. The reason for the time dependency of the strength is the fact that the added load is also time dependent (see Figure 4.4).

A disadvantage turned into an advantage

An interesting challenge in dealing with creative ideas is the turning of the problem into the solution (in the case of the antenna, the source of the problem – ice – is the resource for the solution) or into a new business opportunity (in case of new products). Here too, we may find an opportunity (albeit theoretical) for a new product when we confront the problem: we might manufacture light, mobile yet strong antenna poles for private-individuals in a snowy-icy environment.

If such poles were required by a large enough number of consumers, we could seriously consider the profitability of such a product. Hence, we may consider the Template approach as potential transformation of a problem to opportunity. That opportunity will then be shaped by considering market characteristics (such as preferences and size).

Generalization of the Attribute Dependency Template – an innovative lipstick

If Attribute Dependency could be expressed solely by shifting a static problem into a dynamic state, it could be implemented in only a few cases. The Attribute Dependency Template is much richer, and the following example illustrates the addition of a dependency without relation to the time variable.

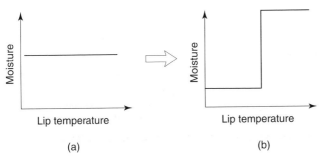

Figure 4.5   Attribute Dependency in lipstick. (a) Common moisturizing lipstick. (b) Lipstick with moisturizing components dependent on lip temperature.

Certain lipsticks contain a capsule of moisturizer (a substance with a high degree of moisture). These capsules burst when the temperature of the lips reaches a certain level. What is the benefit of this product? It has been found that the temperature of lips rises as they dry up. The lipstick has the characteristic of moistening the lips at exactly the right time. Regular lipstick also contains a moisturizing agent, but it evaporates and is "wasted" when the lips do not require it. The innovative lipstick utilizes the moisturizing elements contained in it in the best way. In this case, the dependent variable is the moisture, and the independent variable is the temperature of the lips (see Figure 4.5).

We have now considered several different examples of innovations that share the same structure of a new dependency between two variables: the lighthouse solution, the collapsing antenna pole, the Polo Harlequin®, the Domino pizza service and the lipstick. At this stage we may assume that this structure is a recurring theme that may be exploited to predict future innovations.

Attribute Dependency in nature

In nature, we also encounter many cases that may be described by Attribute Dependency. The chameleon is a striking example. In contrast to other creatures whose skin color does not change, the chameleon's skin color is dependent on its physical environment, thus enabling it to camouflage itself against any background. This attribute of evolutionary adaptability is crucial when the chameleon goes hunting or when it defends itself against enemies.

The basic principle of Attribute Dependency

The uniqueness of Attribute Dependency lies in the possibility to use it for every product or service, without constraints such as the state of the market,

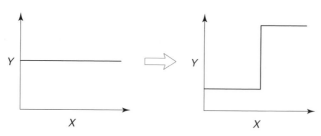

Figure 4.6   Attribute Dependency.

the maturity of the product, or esoteric characteristics defining it or the process of its production. The range of implementation of the Attribute Dependency Template is broad, and in using it one can initiate improvements in existing products or develop ideas for new products.

In the Lighthouse of Alexandria, the antenna and the innovative lipstick, the standard concept of no dependency between two variables was changed to a new concept of interdependency between them. Similarly, the Polo Harlequin® has the characteristic of an Attribute Dependency that makes a new dependency between the type (or location) of a component and its color.

Attribute Dependency is based on finding two independent variables (i.e., a change in one does not cause a change in the other) and creating a new dependency between them.

On the basis of this definition, we may graphically represent a functional connection between two variables, as shown in Figure 4.6. This representation is general, fitting all cases in which a dependency is added.

A graph describing two independent variables has a slope other than 0 ($Y \neq$ constant). The typical graphical depiction of **Attribute Dependency** is: the transition from a straight line graph (no dependency between the variables) to a step function. Mathematically, the function will be described as a split function: $Y_1 = $ constant $Y_1 = Y(X)$.

In order to understand the Attribute Dependency Template more fully, let us look again at the example of a chair and its legs, seat, backrest, armrests, etc. Is the creation of a dependency between the legs and the backrest an attribute dependency? Can a new product be derived from such a dependency?

Let us assume that the answer is positive, and a chair in which the backrest is directly connected to the legs is a new product. Yet, even if this product has many marketing possibilities, it will still not be an example of Attribute Dependency. Attribute Dependency is, for example, the linking of the **length**

Table 4.1 Examples of components and variables

Components	Variables
Eyes	Color of eyes, sharpness of vision
Sugar in a cake	The weight of sugar added in the cake mix, the sweetness of cake
Handle and head of hammer	Length, thickness, height and weight of hammer
Screws	Number of screws, length and thickness of screw, size of screw thread
Alcoholic drink	Percentage or amount of alcohol, color of drink
Hat	Size, color, water-repellence
– – –	Time

of the backrest to the **length** of the legs, which would determine that the longer the backrest, the longer the legs.

The difference between these two examples lies in the definition of the characteristics in the new dependency. In the first case (connection between backrest and legs), the characteristics are the **components** of the chair. In the second case (dependency between the length of the backrest and the length of the legs) the characteristics are the **variables** of the chair. The definition of the term "variable" is not simple. A variable is an element that is subject to measurable change. The measure may be exact (quantitative) or categorical (e.g., the variable of "beauty:" the configuration of the product may be beautiful or repulsive; different taste categories enable us to define beauty as a variable). The legs of the chair are not variables, but their length or color are – since they may vary. Likewise, the moisture in lipstick is a component; the concentration of moisture is a variable.

Table 4.1 presents some examples to clarify the difference between components and variables.

Time is a relevant variable in any product or service, and is always worth including in the list of variables (recall the crucial role of time in the Lighthouse of Alexandria and the antenna pole for the icy conditions). The habit of conceiving systems as static may often cause a fixation of the mind, from which we may extract ourselves with the aid of adding a dependency with time.

In order to understand the main idea of the Attribute Dependency Template and to operate it in connection with the marketing world, let us examine in more detail a hypothetical problem faced by a pizza chain trying to penetrate the market of pizza delivery.

How to compete with "Domino's Pizza" – a hypothetical case

Domino's Pizza is a major player in the world pizza market. This chain based its success on the promise to reduce the price of the pizza if delivery exceeds 30 minutes after the ordering time. This new and original promise, in addition to streamlining the chain, brought it a boom in business, making it a leading world chain for pizza deliveries. It made the consumers aware of the firms' obligation to provide fast service, and also generated an interesting gambling benefit in delivery performance: if a pizza has not arrived within 20 minutes, the customer may hope that it will be further delayed and the price reduced accordingly. While hoping for a further delay the customer is less sensitive to delays.

Considerations of location of sales-points have been emphasized during the firm's growth, so that large areas could be covered in the radius of a 30-minute drive. The logistic setup of Domino's Pizza on such a large scale does not leave room for a major competitor, and competition exists on a neighborhood basis only.

In this situation, two worrying considerations face the potential competitor: his own low ability to penetrate the market and the high ability of the existing chain (Domino's) to overcome him. It is plausible to assume that Domino's will want to remove any new competitor encroaching on its territory as quickly as possible. Ignoring the competitor may promote a situation in which Domino's will have to share the market, possibly leading to a price war and potential long term losses. There may also be other scenarios, such as great improvements in the competitor's business allowing him to become the market leader. In any case, waiting too long may lead to facing a strong and well-based competitor, who fights back relentlessly and poses a real threat to the profits and to the future of the business.

Whether we are dealing with a neighborhood or a country, the considerations are the same. Domino's Pizza will be determined to prevent a small firm from penetrating the market peacefully. It will invest money and effort, and will even be ready to take temporary financial losses in order to drive the new competitor out of the arena. The history of market wars is full of cases in which market leaders bought out competing firms or priced them out of the market, as well as cases where ignoring a small market penetration proved to be a disaster, the intruder becoming a dangerous competitor over time.

We conclude that a "conventional" idea for penetrating the market by a new enterprise is not practical. For example, if the competing chain lowers prices, Domino's will drop prices still lower; if it offers a gift, Domino's will offer a

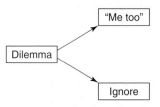

Figure 4.7   Horns of the dilemma facing Domino's Pizza.

better one; if it announces a shorter delivery time, Domino's will react sharply to the threat to its uniqueness and lower delivery time still further, stretching its professional capacity (which is greater than that of an inexperienced competitor). Reis and Trout [1] claim that in such cases the market leader develops a line of defense that is very hard for new competitors to break unless they are equipped with resources and abilities greater than those of the leader. Such marketing wars are well known in market segments in which one firm is dominant.

Does the marketing situation described above prevent anyone from entering the market of pizza delivery? Overcoming Domino's dominance requires sophistication. In this case, a possible penetration strategy is to present Domino's with a dilemma in which it will have to choose between two bad alternatives. Psychological studies of decision-making show that when a director has two bad alternatives he hesitates to choose either one. This hesitation is important to the intruder, allowing him to gain valuable time. Another phenomenon of choosing between two bad alternatives is "the march of folly:" in many cases instead of choosing one bad alternative a compromise is made between the two alternatives, resulting in a path that contains the worst aspects of both.

When the competitor brings an innovation to the concept of service, a dilemma will be created that fits our case: Domino's will have to choose between a move of joining it ("me too") or ignoring it completely (see Figure 4.7). Both alternatives are bad: copying a creative move of the intruder will enable him to conduct a publicity drive emphasizing that it is he who dictates the innovations in the pizza delivery service; ignoring it will allow him to establish himself in the market as its professional leader.

How can a new idea be found that will present Domino's with a dilemma? We must search for a clue to the answer in historical evidence. The meteoric leap of Domino's Pizza was not accidental. It was based on a new and unique approach to the delivery market. This approach must have presented the controlling chains at the time with a dilemma.

Table 4.2 Attribute Dependency in pizza delivery

Variable	Can a dependency be added using this variable?	Reason
Size of pizza	No	Price already depends on size of pizza
Number of extras	No	Price is already dependent on number of extras
Taste	Maybe	A highly relevant variable, but difficult to measure
Adding a drink	No	This is a component and not a variable
Temperature	Yes	Highly relevant variable, easily measured
Distance between customer's home and pizza company	Yes	A measurable variable, but of doubtful relevance
Number of past orders	Maybe	Legitimate and interesting variable, but there is often already a dependency between price and customer habits

Even before Domino's appeared on the scene there were free deliveries, including a promise of fast delivery and tasty pizzas. What was then the novelty offered by this firm? Before it penetrated the market, the price did not depend on the time of delivery (a static price). We have already discussed the fact that Domino's created a new dependency, in which the price was constructed like a step-function – if delivery took up to 30 minutes delivery the full price was charged, and from then on it became cheaper. This is a classic example of the Attribute Dependency Template, in which the independent variable is time, and the dependent variable is the price.

Domino's Pizza methods to control the pizza delivery market can be shown to fit the Attribute Dependency Template. Can one add other dependencies to the delivery service? Will they cause further revolutions in the pizza delivery market? In order not to move too far away from market reality, and in order to facilitate the thought exercise, let us leave the price as the dependent variable (the rationale of such a decision will be explained later in this chapter). Table 4.2 represents a typical thought process for adding a dependency. We screen variables, trying to find good candidates for a new dependency. The following variables appeared in 90 percent of the experimental groups whom we asked to make a list of variables relevant to the problem. For learning purposes, we have also included a few common errors.

Let us look at the temperature as an independent variable. Figure 4.8 shows

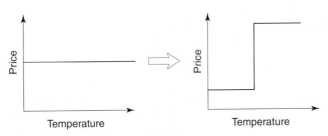

Figure 4.8   Attribute Dependency – the price is temperature dependent.

the following Attribute Dependency: A full price will be charged for a pizza delivered at a certain temperature; below this temperature the price will drop, while holding to the concept of a gamble adopted by Domino's. The marketing message of the new chain will focus on the claim that the pizza's taste depends on its temperature and not on the time of delivery. The time taken for the pizza to arrive does not ensure its temperature, while it is the temperature that is important for the customer. The penetrating chain will promise that if the pizza arrives below a certain temperature the customer will not pay, even if its arrival time is reasonable.

In most of the focus groups that studied this example an argument arose between its supporters and detractors who saw some errors in the concept of "temperature-dependent price." Let us clarify the method of examining the quality of a new idea being considered. We must first ascertain whether we have achieved our goal – has Domino's been presented with a dilemma? A pizza served hot is tastier, and this trait should be primary for a large segment of consumers. The concept suggests that the delivery person will check the temperature of the pizza in front of the customer. If the chain has fulfilled its promise the customer will pay the full price, and if not he or she will be awarded appropriate compensation. The new ritual accompanying delivery affords the new chain a unique opportunity to relinquish the traditional price war and the war for cutting delivery time. The customer's attention may be drawn away from the accepted promise to receive the pizza within 30 minutes, to an equally important question: is it still warm and tasty? The campaign for this move will emphasize the primary importance of taste in the delivery formula, and will announce a focus on measuring the quality of service. Of 200 marketing experts presented with this idea, about 75 percent felt that this concept presents Domino's with a difficult dilemma: to join the "new wave" or to sit idly and see what comes of it.

Now that it is clear that the idea has value, we can examine the difficulties in its implementation. Every new idea entails difficulties and risks. A correct

Table 4.3 "Hot pizza" – turning problems into prospects

Problems	Direction of solution – turning problems into prospect
How to measure temperature in customer's home? (Sticking thermometer into pizza looks unhygienic)	Thermometer must not stick into cheese; temperature may be measured in dough below. A changing-color thermometer will lead to interesting dynamics with customer
Will customer agree to have a multi-use thermometer touch his or her pizza?	A color-changing personal thermometer may be presented to customer as a sales pitch before the declaration. It may be incorporated into magnets sent to customers (with name and phone number for orders)
How can pizzas be kept warm?	If a method is found to keep pizza hot, feasibility and advanced market studies must be conducted. If this question is not answered, the idea is of no use. (An original idea will be presented later)

process of decision-making requires estimating the risks against the potential of the idea. However, it is important to consider each difficulty separately and check whether it is soluble, and not to lump all difficulties into one unit. As we follow the process, we must remember the lesson learned from the example of the antenna in a cold climate: some problems may turn to new prospects. Table 4.3 summarizes the main problems of the idea presented above as they arose in a focus group, and points out ways of turning them to prospects.

The last problem seems to be crucial. If we do not find a solution the thought process we invested in will end. In these cases, we encounter situations in which individuals tend to get frustrated. However we should recall that this is a different problem than the one we started to tackle. Since we occasionally solve problems we should continue to search for a solution to the new problem. In fact it does not necessarily have to be a creative one! Further, if we do find a solution, an opportunity to bake the pizza during the delivery may emerge, leading to shorter delivery time. We will complete the solution of this problem in the next chapter.

Attribute Dependency has created a novel and interesting idea. Like every new idea, it has its drawbacks and uncertainties. It is sometimes possible to overcome these by means of ordinary methods, and sometimes it may be possible to solve a seemingly insoluble problem by using Creativity Templates.

In the case of the pizza delivery we analyzed a marketing problem that can be solved using the Attribute Dependency Template. This is a well-defined

Figure 4.9 A burning candle.

problem. An ill-defined problem is a situation in which a problem is not clearly identified making it difficult to define. The following example will illustrate the interplay between these two situations.

Making a better candle

Looking at a candle, we may notice that this is an interesting and sophisticated system (Figure 4.9). The solid wax serves as fuel for the candle. Without wax, the wick would burn for a few seconds, but if we remove the wick and try to ignite the wax, it would not burn. The wick is, therefore, essential for the fire. The following is the principle of a candle "operation:"

1. The fire melts the wax at the top of the candle to liquid.
2. The wick conducts the liquid wax up by capillary action, bringing it close to the fire. The wax evaporates in the proximity of the fire.
3. A mixture of wax vapor and oxygen in the right proportions is created, which burns and feeds the fire.

In the past, most candles were made of liquid fuel in a dish in which a wick was immersed. How did wax evolve into solid form? Here are two possible scenarios.

1. Market forces projected the need for a tall candle, or for avoiding the awkwardness of an oil dish.

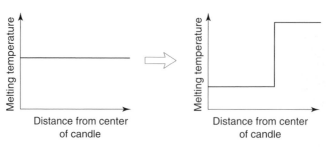

Figure 4.10   Attribute Dependency in a candle.

2. The Replacement Template served as the basis for a thought process in which it was decided to forego the dish and use the oil as its own container and as the carrier of the wick. In order to do this, the state of the oil had to be altered.

We do not need to examine how the candle actually came about; it will suffice to argue that it developed according to certain rules, from which we may receive hints about new candle products. The fact that an essential step in its development may be explained by means of Creativity Templates will encourage us to develop it further by Templates such as Attribute Dependency.

Assume that you are the director of a large candle factory. One morning, the production engineer reports a strange accident. In the process of production, the melting temperature of the outer layer of wax round the candle became higher than that of the inner wax. What will you suggest? As a professional candle-maker, you will probably try to minimize the damage. Yet, as a person aware of Creativity Templates in general and Attribute Dependency in particular, you may discern the Template of Attribute Dependency in this accident. Whereas before the accident the melting temperature of the wax was constant (i.e., it did not change along the radius of the candle), it now changes: as the radius increases, the melting temperature increases. This change is illustrated in Figure 4.10.

Will this accidental event generate a profitable creative idea? After identifying the new configuration of a candle as typical of a Creative Template, we may scan the area of needs and benefits, and try to define a need that may be fulfilled by this change.

How will such a candle burn? As the inner wax melts more quickly, a depression will soon form in the center of the candle (see Figure 4.11). Is there some benefit to this new form of burning? When this attribute dependency was presented to 20 different groups, each group took, on average, 10 minutes to formulate four main benefits:

Figure 4.11   The mode of burning of the candle after addition of dependency.

1. Such a candle does not leak. "Clean candles" may be offered for candle-sticks, cakes, etc.
2. The flame is better protected from gusts of wind.
3 The candle is more economical, as no wax leaks from it.
4. Such a candle has various esthetic and design possibilities.

Multi-colored candles similar to the one described are in fact already being marketed.

Are accidents necessary for locating ideas for new products?

In the candle story we initiated a virtual accident. In real life, such an accident is not necessary for identifying possibilities for Attribute Dependency – we can add a dependency without waiting for unexpected accidents to happen. There is in fact a tendency to believe that chance is a supportive environment for new ideas, and there are many stories about scientific discoveries and new products born out of an unusual event. However, was chance really crucial to their discovery? If we find that these events may be described by means of Attribute Dependency or other Creativity Templates, we may infer that we do not need chance for initiating them – we may well arrive at them by systematic thinking. Moreover, the effectiveness of chance as a means for creative breakthroughs is doubtful. We may remember the few cases that led to novel ideas, but we must also recall all the millions of accidents and chance events that did not bring about any advance in thinking. It is our contention that those accidents and random events that create a new Template matching product configuration manifest creativity and success potential, while those accidents that present generic, non-template based structure will be abandoned.

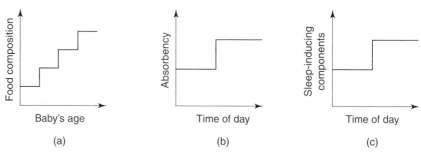

Figure 4.12   Graphical representation of Attribute Dependency added to the product. (a) Baby-food for different development stages. (b) Day-and-night-diapers. (c) Day and night medicines.

Attribute Dependency – between attributes vs. within attributes dependency

Most Attribute Dependencies discussed so far have had a common character. In order to enhance our understanding of Attribute Dependency and understand this distinction, let us consider the following three products:

1. Pain reliever medicaments for treatment of the common cold have been split into day and night medicine: the night medicine contains added ingredients inducing sleep, whereas the day medicine does not.
2. Some brands of disposable diapers are divided into day- and night-diapers. The absorbency and comfort needs are different at night than in the daytime, when the diaper must afford mobility and its absorbency may be smaller because the frequency of change is higher. The parent can select the proper diaper for the time of day.
3. Baby-food was once uniform for all ages, but at some stage different combinations were developed for the different stages of infant development (such as the Gerber® or milk substations stages).

In order to describe the Attribute Dependency added to the product, we shall use the familiar graphic representation with a slight change: the function values will not be continuous. The consumer chooses the value suitable for him, and the function describing a single product is characterized only by the chosen value. The complex graph of Attribute Dependency is therefore composed of segments of different functions, as may be seen in Figure 4.12

The three examples provided above may be analyzed using the concept of Attribute Dependency; but is it similar to the Attribute Dependency we have seen up to now? In the present cases, the addition of a dependency creates different physical versions of the same product, and the consumer may choose the one that suits his or her needs. In the earlier examples, however, the dependency between the two variables was internal, and the "versions" of the product coexisted in the same product. You may recall the varying strength of the pole as dependent on the load placed on the pole: when added strength was needed, the pole "strengthened itself" without the client actively choosing between independent systems. In the case of night- and day-diapers, the parent is presented with two different products and has to choose between them. Such an Attribute Dependency, creating different versions of the product, is called introducing a dependency **between** product attributes, as against introducing a dependency **within** product attributes.

Cycles of dependencies

Most of the previous examples consisted of one step of attribute dependencies, however sometimes a more complex dependency is manifested in a product configuration. Consider a new patent (1999) for a cough syrup: The liquid phase in the bottle allows the syrup to be poured into the spoon easily, but then the syrup phase changes (by itself) to gel to prevent spilling. In the mouth it turns back to liquid, for convenience in swallowing and treatment.

Summary

Attribute Dependency connects two variables not previously connected. The connection is a functional dependency that may be described as a step-function: the dependency between the two independent variables is created at a certain point at which it is needed. Although the logical principle is clear, it is difficult at this stage to apply it to different products or services. How can we locate the variables that will be part of an Attribute Dependency, and how may we assess the feasibility and profitability of the new idea? In the next chapter we shall propose a method for scanning all the information that exists in the product, and for searching methodically and effectively for the potential Attribute Dependencies embedded in it.

REFERENCES

1. Reis, A. and Trout, J. (1992) *Bottom Up Marketing.* New York: McGraw-Hill.

5 The Forecasting Matrix

Searching for Attribute Dependency

We have seen that the Attribute Dependency Template may serve to enable the development of a new product on the basis of an existing one – a standard candle becomes a non-drip candle, lipstick is improved, a new model of car emerges. However, each time we added a dependency and obtained a new product, we did not ask how we might identify the relevant variables, but took these for granted.

In this chapter we shall focus on the methodical identification of these dependencies, and describe in detail a structured process of the "management" of a search for new dependencies, by discussing two examples. In conclusion, we give a systematic operational recipe guiding the practical implementation of the method.

Classification of variables

Our aim is to "read" the information embedded in the product as if we were reading a map of potential changes, and to manage this information in a way that will enable the addition of a dependency. First, we must classify the variables according to the following criteria.

Internal versus external variables

An internal variable is under the full control of the manufacturer, who determines its values and characteristics. We have met internal variables such as the strength of a pole, the price of pizza, the color of a car and the melting temperature of wax.

Table 5.1 Examples of internal and external variables

Product	Internal variables	External variables
Voltaren® (salve to relieve back pain)	Amount of active ingredient, color, smell, fattiness, viscosity	Age and sex of user, time of day, month of year, level of back pain, location of back pain
Television set	Size of screen, volume of loudspeakers, resolution, color of box	Distance between set and viewer, time of use, number of viewers, type of video recorder
Savings plan in bank	Interest rate, exit stations, maximum duration of plan	Size of deposit, income level of depositor, timing of deposit, the building index, rate of inflation, interest in the market
Breakfast in a coffeehouse	Level of spiciness, amount of juice in glass, variety of menu	Outside temperature, number of clients, taste of clients
Computer desk	Type of wood, color, height, surface	Height of user, size of computer, distance from electric outlet

An external variable is a variable of a component within the immediate environment of the product and in direct contact with it, but not under the manufacturer's control. The manufacturer may recommend appropriate values for the external variable, but cannot **ascertain** the achievement of these values (e.g., the manufacturer may not recommend storing a product in the refrigerator, but the consumer may do so). External variables we have encountered so far are time, environmental temperature, load of snow and temperature of lips – all outside the manufacturer's control.

Table 5.1 lists some additional examples of internal and external variables of certain products.

We may observe in this table two phenomena enhancing the search for variables:

1. It is easy to collect external variables from a list of *external components* (those that are not produced by the manufacturer). These external variables come in direct physical contact with the product at a certain time and place.
2. The manufacturer knows the internal variables, and their number is limited.

We may deduce from these observations that the amount of information needed to searching for an added dependency is finite, and it is therefore possible to develop a method to aid us in managing this information.

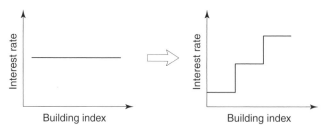

Figure 5.1   Attribute Dependency – a savings plan linked to the building index or rate of inflation.

Dependent and independent variables

Attribute Dependency is the linking of two formerly independent variables. In the past, it was not customary to link the interest rate of savings plans to the building index or rate of inflation. As soon as a bank offered this possibility, a new dependency was added to savings plans. This new dependency may be described graphically (Figure 5.1) in a way that the variable "interest rate" will be dependent on the variable "building index." So far, we have used our intuition to determine whether a variable is on the y or the x axis of the graphs; but if we adhere to the usual definition of the a function, then the variable y is dependent on variable x. The dependency between x and y is not reciprocal: a change in x necessarily causes a change in y, but not the reverse. In the savings plan dependent on the building index, the interest rate changes with changes in the building index. It is impossible to see a practical product in the reverse function: the building index will not change following a change in the interest rate of a savings plan.

1. "A dependent variable" is defined as a variable on the y axis (i.e., the rate of linkage in a savings plan, the price of pizza, the strength of the pole).
2. An "independent variable" is defined as a variable on the x axis (building index, temperature of pizza, load of snow).
3. External variables are by definition not controllable, therefore cannot be dependent variables. Thus, dependent variables will only be internal variables, and independent variables may be either internal or external variables.

The forecasting matrix

In order to enhance the search for dependencies between variables, it is recommended that the product is analyzed using a matrix. The matrix is a tool for systematic scanning of variables, making it possible to predict developments in the product. A forecasting matrix is composed of columns and rows. The columns represent dependent (internal) variables, and the rows represent independent variables (internal or external).

In order to explain the idea of a forecasting matrix and its use, let us take a simple product seemingly lacking any "creative spark" – a drinking glass. We chose this product for its simplicity and familiarity, hoping to convince you that there are no unglamorous products; the very concept of an unglamorous product is in itself unglamorous.

Suppose we are living in a hypothetical world in which all drinking glasses are cylindrical (of constant diameter) and similar in shape and color. There have never been multi-colored glasses or glasses of a complex construction. We, as glass manufacturers, wish to examine new ideas for glasses that will yield new benefits and generate a demand in the market.

Here is a list of some variables relevant to drinking glasses:
1. Typical internal variables: height, diameter, color, heat conduction, transparency.
2. Typical external variables: temperature of drink, outside temperature, level of sugar or alcohol in drink.

Table 5.2 is in the form of a matrix of internal and external variables of a drinking glass. Note that all external variables were chosen from one external component of a glass – the contents (the drink); for the purpose of this example, we have disregarded other external components (e.g., the hands or lips of user or the table).

Each cell in the matrix is called a *matrix element*. We have marked all the elements that we will deal with in this example. Thus, the element belonging to the column *Diameter* and the row *Height* is marked by the letter A. Let us focus on this element. We have assumed that only cylindrical glasses of constant diameter are available on the market. There can therefore be no dependency between the diameter and the height of the glass. The graph describing the dependency between diameter and height of the glass will have a slope of 0, representing no dependency of the dependent variable on the independent variable (y = constant). From our point of view, this is a good starting point for the two variables, since it allows the creation of a new dependency between them, i.e., the addition of a dependency.

Note that there is a single connection between any pair of variables for the two internal variables; i.e., the cell representing the height as dependent on the diameter may also be expressed by the cell representing the diameter as dependent on the height. Even though these are different functions, they have a mathematical dependency (opposite functions), therefore we choose to represent them in the above analysis by one cell only. Each cell linking two internal variables thus represents a function linking the one to the other, as well as the reverse function.

Table 5.2 Forecasting matrix for cylindrical glass

	Height	Diameter	Color	Heat conductivity	Transparency
Height	✕	0 A	0 B	0 C	0 D
Diameter	0 A	✕	0 E	0 F	0 G
Color	0 B	0 E	✕	0 H	1 O
Heat conductivity	0 C	0 F	0 H	✕	0 J
Transparency	0 D	0 G	1 I	0 J	✕
Temperature	0 K	0 L	0 M	0 N	0 O
% alcohol in drink	0	0	0	0	0
% sugar in drink	0	0	0	0	0

Let us consider the first matrix element, in which the height appears as a column variable and a row variable. It is logically ridiculous to discuss a dependency of the height of the glass on its height, just as ridiculous to speak of dependency of the color of the glass on its color. All the elements in which the column variable is identical to the row variable are removed from the discussion, and marked X. Note that the same problem does not exist when the row variable is external, because it is impossible to have external column variables.

We will mark a mode in which there is a prior dependency of two variables by the digit 1. Thus, a black-colored glass is less transparent than a light-gray-colored glass. This mode is represented in matrix element I.

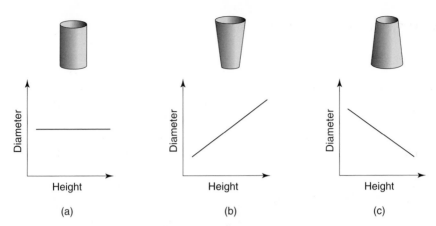

Figure 5.2　Attribute dependency in a glass. (a) Diameter independent of height. (b) Added dependency. (c) Added dependency.

According to this symbolism one may define Attribute Dependency as entailing a change in the value of the cell from 0 to 1.

Elements connected to internal variables

A possible attribute addition to element A in the matrix would be a glass whose diameter changes depending on its height. Figure 5.2a shows a standard glass – the point of departure of searching for new dependencies in the forecasting matrix. Figure 5.2b shows a glass constructed in the form of a frustum. We, the marketing experts for these glasses (hypothetically speaking), must now ask ourselves what benefits are inherent in such a glass. Theoretically, it may be useful to convene "heavy users" and ask them directly what benefit they may derive from this glass. In this case, the answer would be clear enough: the storage space needed for such a glass is much smaller than that for a cylindrical one, because such glasses fit into each other like empty ice cream cones. This is the glass's main benefit, and one that is utilized particularly in disposable glasses.

Figure 5.2c represents another possible dependency. Here, too, the glass is frustum-shaped, but its slant is in the opposite direction – the diameter is largest at its base. This is also an added dependency, for which a benefit might be found. The odd shape of the glass has an advantage for use in an unstable environment, such as a train or a car, or for people lying in bed such as patients in hospital: the reversed slant prevents the liquid from spilling and increases the stability of the glass. Another benefit is that the surface of the liquid in

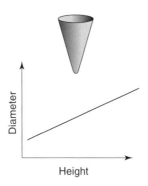

Figure 5.3   A glass that does not stand.

touch with the outside air is smaller, thus keeping the liquid in the glass warm or cold for a longer time.

Before we proceed to the other elements, let us emphasize the importance of conducting a thorough investigation of every element in the matrix before we conclude that "there is no potential for an added dependency," or that "the existing definitions of the new dependencies are sufficient." If we chose to add a dependency as shown in Figure 5.3, we would have a cone-shaped glass pointed at the bottom, not able to stand at all. Yet even this idea, far-fetched as it may seem, must not be discarded at the outset. In fact, such a glass was in use many years ago: the High Priest at the Holy Temple held it in his hand with the blood of the offerings. Since it could not be put down, the priest had to hold it during the whole ceremony. This had an important effect: the movements and warmth of his hand kept the blood from clotting. The disadvantage of the glass became its advantage in this special case. Are there other, less specific cases in which an unstable glass would be an advantage? If we proceed forward in history, we will find that the interest in a glass that does not stand has not disappeared. Edward De Bono in his lectures proposed such a glass for cocktail parties. After the guests finish drinking, they look for a place to put the glass down without it toppling over. They will find a special tray in the corner of the room, with holes accommodating these glasses. The guests, wishing to avoid the unpleasantness of leaving their glasses "lying" all over the room, will put them in the tray and the labor of collecting the glasses will be saved[1]. We also note that in some bars drinks are served in such glasses, to increase consumption: since they cannot be put down, drinking volume increases.

---

[1] The cleaning problems that can be imagined if the guests did not put the glasses in the tray may explain why such a product does not exist.

This is not the end of the story for the "glass that cannot stand." In courses we conduct, the participants have circumvented the problem of making the glass stand by finding an environment in which it stands better than in others – at the seashore. We can make use of this glass at the seashore because it is easy to stick it in the sand, and its stability is then greater than that of a normal glass. This is another example of a disadvantage (instability) that becomes an advantage (more stability than a normal glass).

Function Follows Form (FFF)

In the above example the specific new attribute dependency was first defined, and only later was consideration given to its practical applications in the market. A thought process in which the new configuration is first defined and only then is scanning conducted for possible benefits that can be derived from it, sounds surprising and innovative. According to this method, we find a solution and then locate a problem to fit it – in contrast to the usual process in which we first define a problem and then search for its solution.

There is a second principle of creativity on which the Creativity Templates approach depends. It is termed "Function Follows Form." Also suggested recently in cognitive psychology [1, 2], this principle states that it is easier to find functions or uses for a certain form or structure once the form is generated, than the other way around. This indicates an "asymmetry" in thought: people are capable of achieving better results when a form is already defined, that is when a pre-inventive structure exists.

The Function Follows Form principle may best be demonstrated by an example. It is straightforward to compute the product of 59 and 83, but difficult to find the factors of 4897. Another example is trying to find the angle to point a gun for a bullet to fall at a given distance. The distance can of course be determined by simply firing and measuring where the bullets falls. It is much more difficult to find the angle at which to point the gun so that the bullet will fall at a given distance. Usually, this inverse situation needs much trial and error to resolve, and it is rarely precise.

This is where Creativity Templates come in. By altering the configuration of a product to a pre-invented structure, the Creativity Templates provide efficient structures for the conception of a new form. All that follows is the easier task of searching for an appropriate function.

In their research, Finke, Ward and Smith ([1] noticed the existence of two distinct processing phases of inventive thinking: a generative phase, followed by an exploratory phase. In the initial, generative phase, people construct mental representations (pre-inventive structures) when they process various properties that promote creative discovery. These properties are then exploited in the exploratory phase, in which people seek to interpret the pre-inventive structures in meaningful ways. This sequence of events underlies the FFF principle. Accordingly, people are more likely to make creative discoveries when they analyze novel forms and then assess the benefits they may project, than by trying to create the optimal form based solely on the desired benefits identified in an analysis of market demand.

These findings about the asymmetry of human thinking support the following approach: If we wish to arrive at a creative idea for a new product, we must first turn to the information embedded in existing product (configurations), and only then to the information available in the market (benefits, needs). In a recent study, [3, 4] evidence was found in support of the real contribution of the FFF principle to the success of product ideas.

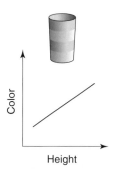

Color

Height

Figure 5.4   A glass with varying color.

So far we have only dealt with one element of the whole matrix. The matrix enables us to search for additional dependency by a systematic scanning of all the elements, in order to find 0 modes and change them to 1 modes. Such scanning enables us to extract the maximum number of creative ideas that may be yielded by an existing product.

In the next element (B), the dependent variable is the color and the independent one is the height. In the initial mode, the color of the glass was uniform. There may be colored glasses, but their color is constant throughout their height. We have already seen color as dependent on another variable, in the Polo Harlequin®. The variations in color usually create social benefits, as they often indicate daring or an esthetic sense.

Although a colored drinking glass may indicate something about the esthetic sense of the consumer, we can use Attribute Dependency to examine whether a glass with a dark-coloured bottom and a light-coloured top (see Figure 5.4) can have any practical benefit.

We can actually think of a benefit in changing the color of the glass. Many juices contain matter that has not been well dissolved such as fruit particles or edible dyes (e.g., home-squeezed orange juice). These solid components are not noticed in drinking, and are part of the juice's taste and consistency; however, the sight of the residue can be unpleasant. By playing with the shades of color of the glass we may esthetically improve the drinking experience.

Even an idea that seems at first hand to have a low market potential is worthy of fair examination. Maybe this glass does not seem like great news for glass manufacturers, *but what about bottle manufacturers?* An analysis of a similar forecasting matrix may be done by a manufacturer of drink bottles. Here, he or she could provide a significant benefit to a definite public (juice manufacturers who think they might sell more if consumers could not see the residue).

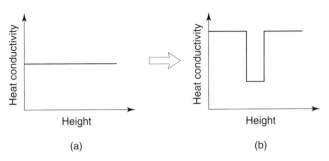

Figure 5.5   A glass with varying heat insulation. (a) Essential Attribute Dependency. (b) New role configuration.

The examination is simple enough. One has to ascertain the proportion of this public, and the rate by which the change will increase its consumption. There are simple procedures, both quantitative and qualitative, for examining these two questions.

Let us follow the first row of the matrix to element C, in which heat conductivity is dependent on height. This element may be described by a step-function, in which the glass will conduct heat at the bottom, but will be insulated at the top (see Figure 5.5). We must now ask when and why we need a glass made of insulated material, and when we need a glass that will conduct heat. The market answer will be that we are not interested in drinking a boiling-hot drink, therefore the glass must conduct the heat outward (i.e., cool it); but we do not want the glass to be too hot to hold, therefore it is important to insulate it.

Manufacturers usually add a handle to the glass in order to solve this dilemma; but by adding a dependency we may think of a glass with insulation strips in the area where it is held, while it conducts heat through the rest of its surface. Such a glass need not have a handle, and so will fit into places where storage space is at a premium (e.g., next to coffee machines).

The graph for describing the new glass is now brought up-to-date: after we have "captured" the idea, the product may be planned and constructed so that it will accommodate the needs of a maximum number of the target population.

This is the time to remind ourselves that the glass serves as a thinking exercise, and the products we derive from the matrix are not necessarily practical for marketing. Even when this process is effected on a real production line, we cannot foretell whether a product will emerge that will entirely change the production line, or whether it will survive merely as an original idea. Let us continue with our analysis, and we may be lucky enough to find a more attractive idea, more relevant for our time.

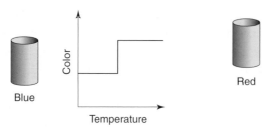

Figure 5.6   Dependency between drink temperature and glass color.

Elements connected with external variables

In matrix element K, the independent variable (the temperature of the drink) is not under the manufacturer's control. It is possible to change the height of the glass as a dependency on temperature in one of two ways:

1. An irreversible change: the glass will become shorter as it heats up, not returning to its original size as it cools.
2. A reversible change: the glass will change in height according to the temperature of the drink in it.

Before we discuss the marketing significance, we should examine whether the added dependency can be practically implemented. While an irreversible change is easier to implement, the reversible change requires material having a "memory of shape by temperature" – such as Nitinol®. Such materials are not expensive, but the development of a glass of varying height may be complicated and costly for most glass manufacturers. Since a cursory look at the market does not point to any real advantage to adding such a dependency, we may decide not to pursue this matrix element. Note that often a change in the product requires the investment of development resources too costly for the firm. In case of a significant difficulty in the technical feasibility of the idea, it is of no use to continue analyzing its possible benefits in the market.

We shall now discuss matrix element M, in which the color of the glass depends on the temperature of the drink served in it. Are we in a 0 mode in this matrix element? Although there are glasses with a design on them which changes upon heating the glass, this is mostly a gimmick. Although this can create social benefits (demonstrating the esthetic sense of the consumer) there is no beneficial dependency that serves the glass directly; many other products could serve as well for this purpose. We will therefore treat this element as a mode of independence, which can be changed to a dependency as shown in Figure 5.6.

Our method requires that we first conduct a technological feasibility study:

are suitable materials available? What are the development costs? etc. In this case, the answer is immediate: such glasses have already been approved for use (and for contact with food). Actually, materials having the characteristic of changing color are not costly, and therefore we may continue examining the market for benefits derived from it.

Who needs such a glass and why? How will they use it, and how much will they be ready to pay for it? Marketers who have dealt with these questions in our studies suggest that the target population will be parents (mainly of a first child). The user will be the child who drinks warm cocoa, who will know by the marking on the glass when it is safe to drink without being scalded. It has been claimed that parents will be willing to pay a higher price for such a glass for its safety value.

At this point we would like to draw attention to the difference between creative and conventional marketing approaches. The fact that we found a new benefit so fast must not discourage us from striving for additional ideas. It may be possible to find other ideas or to improve this one, and in fact several marketing people carried this line of thought further, and arrived at a more attractive idea: the new attribute dependency may be used in baby bottles. The baby-formula is warmed up during preparation. A thermometer may be a quick indicator as to whether the formula is at the proper temperature, or whether we must wait a little longer until it cools. All that is technically required is to add a color-changing thermometer to the baby bottle itself. This added safety attribute may increase the target population as well as the profits.

This template and its generalizability across products can be illustrated with an example drawn from a recent study by Andrews and Smith [5]: Hungry Jack® syrup bottles are designed for use in a microwave oven. The bottle labels change color on reaching a certain temperature, thereby informing consumers that the syrup is ready. This example contains the key features of the same Template. The two independent variables, in this case, are temperature and label color. A dependency is created by a step-function between these two variables. Up to a critical temperature, the label color is not activated, and on reaching it, the color changes. The generalized form no longer involves the specific variables of the earlier examples, nor does it necessarily involve the same product (or service). Yet, it is identifiable and general across products and services and, thus, it is defined as a Template.

Encouraged by the fact that we have succeeded in devising a useful idea by using matrix element M, we shall now turn to matrix element N, defining the dependency between heat conductivity of the glass and the temperature of the drink.

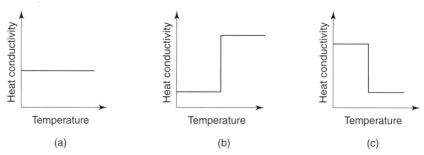

Figure 5.7   Heat conductivity dependence on drink temperature. (a) Normal state. (b) Conductivity increases (insulation decreases) as heat of drink rises. (c) Conductivity decreases (insulation increases) as heat of drink rises. Note that although we have used a step-function for clarity in nature a linear (monotonic) dependence is more realistic.

When we ask ourselves whether there are glasses in which a change in the temperature of the drink causes a change in their conductivity (or insulation), the answer is "no"; all known glasses "behave" in the way indicated in Figure 5.7a.

In principle, an Attribute Dependency may mean one of two variants (see Figure 5.7b, c):

1. Heat conductivity will increase as the drink becomes hotter.
2. Heat conductivity will decrease as the drink becomes hotter.

In the first case, the behavior will be as follows. When the tea or coffee is very hot, the glass will cool them by distributing the heat outward. When the temperature comes down to a certain level (suitable for drinking), the glass will become insulated and conserve the heat. In the second case the behavior will be converse; hot coffee will not exchange heat with the environment, and after it cools, the glass will become heat-conducting and enhance the cooling of the liquid in it. The second idea is naturally not in accord with our drinking habits of hot drinks, and we will not consider it further.

Let us examine the first possibility of product behavior – a direct correlation between the level of conductivity and the drink temperature. Such a glass may solve the problem of drinking coffee in a cold environment, in which hot coffee cools very fast. A glass with varying heat conductivity will enable coffee to cool down quickly to drinking temperature, and delay cooling after that, so that we may enjoy a warm drink for a longer time.

The answer to the question about the feasibility of such materials is quite clear: there are materials available for use for the production of such a glass, but they may raise its price to such a level as will make it difficult to sell. The preferences of the target population and the price it is willing to pay for realizing them can be analyzed by the usual methods of market research. If we find that the higher price is not feasible in view of the expected rate of sale, we must drop the idea.

The example of the drinking glass was chosen in order to demonstrate simply and clearly the principles of using the matrix; but it may have given the impression that most ideas are either trivial or do not have practical value. In order to instill in the readers a more positive feeling about the applicability of this method, we shall now deal with a real-life situation that was handled by means of the forecasting matrix. We shall discuss practical dilemmas that arose in this case, from which we may evaluate the contribution of the forecasting matrix at the stage of searching for new ideas. The following example was taken from [6].

Forecasting matrix – analyzing baby ointment

Description of a marketing state

Let us consider a hypothetical scenario. Assume a large company specializes in cosmetics and pharmaceuticals. The directorate of this company has decided to expand into the market of baby ointments. In order to achieve this goal, the company considers the possibility of launching a product with a new benefit. It does not intend using its reputation in other areas for promoting this new product, therefore the benefits it wants to present need to be meaningful and clear to the consuming public.

Let us assume that we are members of a small team making the decisions about which product to develop and how to introduce it to the market.

Description of the product

Baby ointment is designed to ease the pain from rashes on a baby's delicate skin, to heal the baby's skin and possibly to prevent the rash from reappearing. Rashes appear mostly in the groin area due to prolonged contact with the diapers that absorb the baby's excretions. The ointment under discussion is composed of a fatty substance, a moisturizer for nourishing the skin and an active ingredient for healing the skin.

Basically, baby ointments have not been changed in any significant way since their creation in the early 1900s. There are several brands that can be classified by their viscosity and the concentration of active ingredients and moisturisers. There are also ointments made up by individual pharmacists that are also considered effective, and also have loyal customers.

We can assume that almost any manufacturer can produce any other's product in a similar quality. The differences between brands (e.g., viscosity of A and B) stem from analyzing needs or from various marketing approaches, not from technological constraints. Let us assume further that our firm has a proven research and development potential, so we need not worry about technological inferiority to our competitors.

Many books and academic papers deal with "how to design a product so that it will penetrate the market and prevail in it." There are several methods to examine consumers' preferences (such as conjoint analysis), that may help us to determine the preferred profile of the new product. In this book we propose another approach. This new approach does not replace the existing methods of market analysis and designing the product according to consumers' preferences, but is meant to help us in finding new characteristics not overt in existing products – characteristics that the consumers are not yet aware that they need. Let us emphasize again that we should not neglect the methods for market analysis once the idea has crystallized. On the contrary, in order to design the product so that it will have the greatest chance to penetrate the market, the company must include these methods in the process of development.

Baby ointment matrix

Consumers in a few in-depth interviews noted the following variables:
- **Internal variables:** viscosity of the ointment, odor, amount of fatty substance, color, amount of active substance.
- **External variables:** amount of excretions at a given moment, acidity of excretion, sensitivity of the baby's skin, the baby's age, type of food the baby consumes, time of day.

Once the variables' space is outlined a forecasting matrix is constructed. Again, the columns of the matrix consist solely of internal variables; the rows of the matrix may consist of a mixture of internal and external variables. When there is no dependence between two variables, the relevant matrix element is marked with a zero. There is no dependency between color of ointment and amount of excretions at a given moment (element D1). Thus D1 is marked 0. A partial matrix is presented in Table 5.3

All the elements in the above matrix are in the "zero mode." This denotes

Table 5.3 A partial matrix of baby ointment variables

	Viscosity A	Odor B	Amount of active substance C	Color D	Amount of fatty substance E
Amount of excretions at given moment 1	0	0	0	0	0
Acidity of excretion 2	0	0	0	0	0
Sensitivity of skin 3	0	0	0	0	0
Age of baby 4	0	0	0	0	0
Type of food 5	0	0	0	0	0
Time of day 6	0	0	0	0	0

that no relationship exists between the dependent and independent variables. Does the above matrix indicate anything about the product? the categories? the marketplace? Although the zero-mode matrix is composed of data about the product, it actually reflects the market situation. The ointment, which is considered a high-quality product, has not changed for decades. This lack of change is illustrated by the zero-mode, degenerated, matrix above. This situation may have three possible explanations:

1. **The manufacturer predicted the market accurately a hundred years ahead.** The original designer and planner of the ointment might have understood market needs, changes in associated products such as diapers, and societal and family shifts so well that the company need not search for new products that were unavailable at that time.

2. **The market for ointment has not undergone any social or economic change.** Needs that existed decades ago have been frozen and are still relevant today.

3. **For various reasons, manufacturers have refrained from disturbing the market to search for new needs.** This inactivity has caused a kind of slumber in both market expectation and product development. As awareness of needs arises, product development occurs followed by emergence of new products.

In the complex reality of the marketing world, each of the above explanations theoretically may be true. If we were conducting this analysis in 1910 (a few years after the first ointments were introduced), explanation 2 would be the most likely. Present-day reality, however, clearly points to explanation 3.

Although the manufacturers have not consciously anesthetized the market, there is a reason to believe that we are standing in the quiet before the storm – an avalanche of new products will flood the marketplace as soon as the first new product is introduced.

This example is a rare and extreme case. In most cases the picture will not be so sharp nor will events in the marketplace take such a dramatic turn. Even so, we argue that the matrix can be used to predict new products on the horizon – though not the best timing for their introduction to the market. Timing, launch strategy and appropriate marketing mix are not tied to the methodical search for new product ideas. Instead, they follow the process for identifying products that may seem promising at first glance. The purpose of the matrix is to elicit promising ideas to be introduced when the time is right.

The matrix may help us develop new ideas through the Attribute Dependency Template mentioned above, using the following procedure.

1. For each "zero-mode" element define a new dependency, namely, how will the two independent variables become dependent on each other?
2. Preliminarily examine whether the product is able to sustain the added Attribute Dependency.
3. Find and define the benefits of the new idea by searching the new product structure. What would be the advantage for the customer of using the new product?

Let us review a few elements of the matrix one by one.

Matrix cell B1: Odor and amount of excretion

1. Define the new dependency. In contemporary baby ointment products, the ointment's odor remains constant regardless of the amount of a baby's excretion. We may add a new dependency, whereby the ointment remains odorless as long as no excretions are present in the diaper, but it gives off a (pleasant) fragrance as soon as excrement is present.

2. Ascertain product feasibility. At this stage we must address the question, "Is there any use in continuing in this direction?" We could probably find some way to introduce into the ointment tiny capsules containing a pleasant-smelling substance. The capsules could burst when they come into contact with an acidic substance such as excretion, releasing pleasant odor into the air. However, if the addition of this dependency would require substantive R&D or the addition of toxic substances into the ointment, the idea is likely to be discarded at this point.

3. Search for benefits. One must tread carefully here. At first glance, we may not see the need for an ointment that changes its fragrance. Solid excretion gives off an odor in any case, announcing its appearance. Liquid excretions, which are less odorous, are already marked by a change in the color of the diaper.

Yet the experienced user of Templates will notice a valuable clue: the addition of an excretion-dependent color to the diaper clearly shows both the **need and relevance** for an excretion-dependent odor. Careful examination will reveal the reason. When diapers are covered with layers of clothing, a visual indication of the presence of liquid excretion will not be noticeable. An odorous ointment would provide parents with a distinct advantage, sparing them the need to remove the baby's pants to check on the state of the diaper.

Matrix cell A6: Time and viscosity

1. Define the new dependency. In contemporary baby ointment products, the ointment's viscosity is not time-dependent, having the same viscosity day and night. We may introduce a new dependency, whereby the ointment will be viscous at certain times of the day and liquid at other times.

2. Ascertain product feasibility. At first glance, periodic changes in the viscosity of the ointment may seem too complex and costly to develop, even without considering the benefits of such an ointment. The first reaction may therefore be to rule out this possibility.

Here we would be prudent to note the existence of two different kinds of attribute dependencies: those **within** the product attributes, and those **between** them. In cases where it would be impossible to impose a "spontaneous" alteration of the product, a dependency that the customer can control directly may be added instead (e.g., as in pain relievers for day and night). Accordingly, we may present a package containing two ointments – one thick (viscous) and one light and airy (liquid) – for the parent to use at the prescribed times.

3. Search for benefits. Without a clear advantage, no parent will buy a double quantity of ointment or waste his or her time using different ointments at different times. Clear definition of the timing (i.e., when to use a viscous ointment and when to use a lighter, airier one) is required here.

Answers to a short questionnaire (regarding this new concept) filled out by parents have shown that a viscous ointment is advantageous at night when diapers are changed less frequently, since the ointment can serve as a barrier between the excretions and the baby's sensitive skin. In the daytime, when babies' diapers are changed more frequently, we may allow the baby's skin to "breathe" by using the lighter cream.

Accordingly, the customer might be presented with this new concept while stressing the above-mentioned benefits. Parents would appreciate a viscous ointment at night so that both they and their babies will sleep peacefully, and a lighter liquid ointment in the day, to allow their babies' skin to breathe. When presented with a choice between this adaptive ointment and an ointment with a constant viscosity, the customer may associate the adaptive ointments with other adaptive habits – day and night pain relievers, day- and night-diapers, etc. Thus he or she may agree to try the new product concept.

Column C

Let us now examine the possibilities in column C of the matrix. The concentration of active ingredient in the product is currently the same for all existing ointments. Remembering that the added dependency may be delivered between the product attributes, we will ignore for the moment the difficulty of a "spontaneous" change in the ointment. Let us instead offer a series of ointments with different concentrations of active ingredient. The new dependencies will be expressed by external variables: baby's age, type of diet and degree of skin sensitivity.

Consider the dependency between the concentration of active ingredient and the baby's diet. Newborns usually begin their lives nursing on mother's milk, graduate to a synthetic milk formula, then progress to baby food. They may also receive homemade pureed vegetables or soups. Each dietary stage contributes to a different acidity in their excretions, and thus to differing exposures to skin irritation. Here, the added dependency may be expressed as a series of ointments adapted to each dietary stage. Again, the launched product may be better suited to each dietary stage via different combinations of ingredients – in contrast to an all-circumstance product – thus attracting the attention of a parent. Column C contains several possibilities for such ointments targeting varying circumstances.

Managing the ideation process issues

A degenerated matrix versus a saturated matrix

Baby ointment has been described through a matrix in which all elements were in the zero mode. We may assume that in time many "1" positions will appear on this matrix. Let us define two extremes, a degenerated matrix and a saturated matrix.

- A *degenerated matrix* is one in which all or most elements are in the zero mode (i.e., Figure 5.8a).

	A	B	C	D
1	0	0	0	0
2	0	0	0	0
3	0	0	0	0
4	0	0	0	0

(a)

	A	B	C	D
1	1	1	1	1
2	1	1	1	1
3	1	1	1	1
4	1	1	1	1

(b)

Figure 5.8   A degenerated matrix (a) and a saturated matrix (b).

- A *saturated matrix* is one in which most cells are in mode "1," indicating that many variables are interdependent (Figure 5.8b).

A degenerated matrix suggests there is the potential to offer the market new products with several new benefits that have not yet materialized. The company has to examine the market situation carefully in order to arrive at an appropriate decision. Should the firm introduce new products? Should it enter this market at all? Should it wait, and if so, how long? By contrast, a saturated matrix suggests that the firm may have "missed the boat:" communication between product developers and the market has already yielded many new Attribute Dependency based products. Now that the firm has decided to join the game little room for maneuvering is left.

Coming across a saturated matrix situation, we must realize that analysis will be much more difficult. The number of innovations that our matrix can help us find will be smaller and in many cases may have already been foreseen. A saturated matrix, however, contains other valuable information. The product may indeed have exhausted its potential for development, from this Template's perspective. If so, no marketing efforts will change this fact. In such a case, two alternatives should be examined:

1. **Analysis of another product.** Often, a firm has several products in which changes may be made.
2. **Use of another Template.** The rules of product development obey codes other than Attribute Dependency. In the following chapters we shall present other analyzing tools enabling the innovation of products.

In reality, matrices are rarely extreme, but rather exhibit intermediate situations. Therefore, the decision-making process must be carried out while considering at the same time the extent of the matrix's inclination to one or the other extreme – degeneration or saturation. One should also be aware of the dynamics in the categories of the products under consideration.

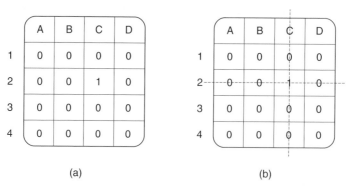

	A	B	C	D
1	0	0	0	0
2	0	0	1	0
3	0	0	0	0
4	0	0	0	0

(a)

	A	B	C	D
1	0	0	0	0
2	0	0	1	0
3	0	0	0	0
4	0	0	0	0

(b)

Figure 5.9   (a) Portion of a matrix. (b) Strategy of columns and rows.

In practice, it is not always necessary to prepare a matrix for each and every product in order to practise observing products or to find new opportunities. The purpose of the matrix is to regulate our thinking by establishing an order with easy-to-follow rules.

Improving scanning efficiency through heuristics

In many cases, the matrix contains 15–20 variables. When we are dealing with a service, and not with a product having physical characteristics, we may have much larger matrices. A systematic scan by column and row takes a short time, relative to the time invested in searching for an idea unsystematically. The information yielded by the analysis of such a matrix is also much more limited than the information derived from a market study; but one may much improve and shorten the time of matrix scanning.

For the sake of convenience, we shall examine a small square portion of a matrix (Figure 5.9a). Except for element C2, all elements are independent (0 mode). This means that someone has succeeded in linking variable C with variable 2, deriving a significant benefit for some market. We may conclude that the dependent variable may be manipulated, and the independent variable 2 may be easily linked to another dependent variable. Moreover, at least one of the two variables is meaningful to the consumer. According to the suggested thinking strategy, we should continue scanning row 2 and column C (Figure 5.9b) before we go on to other elements, because there is a greater chance of adding a significant dependency in these than by randomly choosing other elements of the matrix.

In the matrix for baby ointment, for example (Table 5.3), we created a dependency in matrix element B1. Even if the idea does not mature into a product, we have learned that the dependent variable "odor" may potentially be linked with an independent variable. We shall therefore scan all the possible dependencies between the variable "odor" and independent variables. Moving along the column B of the matrix may be more productive than a random choice of elements. According to the same principle, the variable "amount of excretions" has also "proved" its ability to link with a dependent variable, and therefore it is highly probable that we will find a new dependency for it by moving along row 1 of the matrix.

If the matrix before us is degenerated, we should randomly choose three elements and examine the possibility of adding a dependency to each of them. If we succeed, we should continue following the column strategy to further successes.

In most cases, we will face a matrix having many 0 modes and a smaller number of 1 modes. In this case, we should randomly choose two 0 modes for examining the possibility of adding a dependency, and one matrix element of 1 mode. We now again have three elements as a basis for scanning of rows and columns. This subject is expressed in the operational prescription below.

Summary

The approach described here is a heuristic one. Heuristics is a "rule of thumb," it is sometimes worth using, and sometimes it is better to refrain from using it. This is similar to the rule in chess that one must save the queen for the last stages of the game, and she must definitely not be exchanged for an inferior piece such as a rook. This rule facilitates the game, as it prevents wasting thinking time. Yet, sometimes a chance appears of exchanging the queen for a pawn, in order to achieve a faster checkmate later on. In such a case, the automatic use of heuristics causes a renunciation in advance of the chance to win the game. In using a matrix one should consider the suggested approach, but also keep an open mind and be flexible when necessary.

The forecasting matrix enables us to prepare a reserve of new products, and to plan a strategy for their production and presentation to the market even before the actual marketing step. This "leisurely" preparation enables us to choose modes of action and timing that may, in time, lead to an advantage over our competitors.

Operational prescription

Operational prescription for a forecasting matrix

1. Make a list of internal variables (under the manufacturer's control).
2. Make a list of external variables (not under the manufacturer's control).
3. Build a matrix in which the column variables are composed of internal variables and the row variables of all variables.
4. For each cell mark whether it is in 0 mode (no dependency between elements) or in 1 mode (dependency exists).
5. Examine the tendency of the matrix (degenerated or saturated), and decide whether to continue.
6. Choose three elements and try to add a dependency (change to 1 mode).
7. For each element in which you succeeded in adding a dependency, assess the feasibility of the change from a practical point of view.
8. Search for a new benefit or significance for the market derived from the addition of a dependency. Remember that the principle of Function Follows Form supports you. Do not overlook new market segments that may derive from this addition of a dependency.
9. According to the measure of success, choose a scanning strategy by columns and rows, or choose three additional elements.

REFERENCES

1. Finke, R. A., Ward, T. B. and Smith, S. M. (1992) *Creative Cognition.* Cambridge, MA: MIT Press.
2. Dominowski, L. R. (1995) "Productive Problem Solving," in *Creative Cognition Approach,* S. M. Smith, T. B. Ward and R. A. Finke (eds.) Cambridge, MA: MIT Press, 71–95.
3. Goldenberg, J., Lehmann, R. D. and Mazursky, D. (1999) "The Primacy of the Idea Itself as a Predictor of New Product Success," a Marketing Science Institute working paper, Report No. 99–1105.
4. Goldenberg, J., Lehmann, R. D. and Mazursky, D. (2001) "The idea itself and the circumstances of its emergence as predictors of new product success," *Management Science,* **47**(1), 69–84.
5. Andrews, J. and Smith, D. C. (1996) "In search of the marketing imagination: factors affecting the creativity of marketing programs for mature products," *Journal of Marketing Research,* **33** (May), 174–187.
6. Goldenberg, J. and Mazursky, D. (1999) "The voice of the product: templates of new product emergence," *Innovation and creativity Management,* **8** (3), 157–164.

6 The Replacement Template

To see what is in front of one's nose needs a constant struggle

George Orwell

What is the Replacement Template?

We have already encountered several examples of the Replacement Template, for instance

1. Edison's legendary gate which forced his guests to activate his private water pump.
2. A keyboard of a portable computer which transforms mechanical energy (from the user's fingers) to charge the battery.
3. The Wirefree device which uses the loudspeakers from a car's radio system to improve the sound quality of the cellular phone.
4. Antenna pole in which the ice that accumulated in the environment was used to increase its sturdiness.

One abstract structure surfaces from all of the above ideas, based on their underlying code – harnessing existing resources from the immediate environment to replace a product component which fulfills the same needed function. Another such code may be extracted from the following illustrations.

The Replacement Template creates a link between a resource (material, energy or a phenomenon) existing in the environment and a role that requires fulfillment. Thus, the system saves resources while it becomes more "compact." In most cases two existing components are connected, but we must remember that systems undergoing Replacement are likely to be considered creative even when no resources are saved.

In this chapter we will present the structure of this template in detail and suggest the best way to implement it. To illustrate its generalization let us look at some examples before we provide a complete formulation of the Replacement Template.

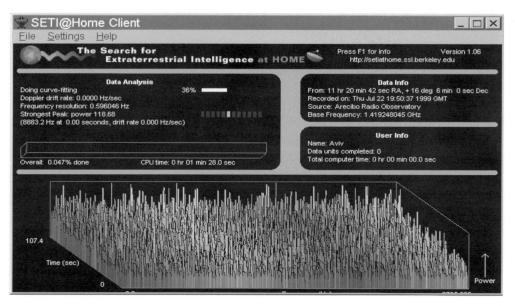

Figure 6.1   SETI Screensaver.

The Search for Extraterrestrial Intelligence (SETI)

SETI is the name of a project that is searching for extraterrestrial intelligence. Radio frequency waves are received, recorded and analyzed to see whether some order can be identified to indicate the existence of intelligent society in space. The required computing resources were much greater than those available to the research workers. A solution to this problem was to harnesses the power of hundreds of thousands of Internet-connected computers to the search for extraterrestrial intelligence. Anyone can participate by running a screensaver-like program on his or her personal computer (PC) that downloads and analyzes radio-telescope data. Thus, the individual PC computation resources are harnessed to the task of analyzing a huge amount of data, which is returned to SETI for further analysis and coding. The screensaver is presented in Figure 6.1. How similar would you consider this structure to that of Edison's gate and the idea of energy extraction in the portable computer? Bear in mind the benefits associated with the close-to-zero costs of adopting these resources in the three cases (unless, of course, some of Edison's guests turned away from a visit, or some PC users are constantly distracted by the SETI graphs on their "screensavers").

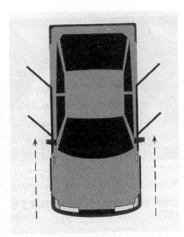

Figure 6.2   Direction of opening of car doors as a safety device, to prevent opening while in motion.

Doors in cars

The door hinges of most cars today are situated at the front (see Figure 6.2). Why does the door of the car not open in the direction of exit of the passenger, so as to make it easier to get in and out? The reason here is safety. The designer wanted to prevent the door opening while the car is in motion, especially at high speed when there is a danger that a passenger might be tossed out. Instead of building a safety device which would shut the door if it opens in mid-drive, the air current is used to apply pressure on the door and prevent it from opening widely. The existing resource is the air pressure.

Replacement in Russian theater

In some theaters in Russia there is no curtain. When the scenery needs to be changed during the performance, the stage is lit by projectors at a certain angle (see Figure 6.3). The reflection of light from the air particles creates a kind of halo, preventing the viewers from seeing the stage. This halo serves as a cheap curtain, since the projectors exist in the theater in any case.

All these ideas which once were novel and surprising, although from remote fields of expertise, share the same general structure that was presented above – the Replacement Template. In order to understand the required thought dynamics and become acquainted with the unique features of this Template let us examine a hypothetical problem

Figure 6.3   A halo of light as a substitute for a theater curtain.

A flat tire in the wrong place and time

Let us examine the following problem: a driver on a deserted road at midnight suddenly notices a puncture in one of the wheels. When she tries to change the wheel, she realizes that the nuts and bolts have rusted to a point where she cannot open them with her tools.

Here are some ideas for solving the problem, along with an assessment of their creativity, gleaned from 20 groups who handled this problem (Table 6.1). The second column in the table marks the proximity between the existing resource and the wheel. The solution of using a mobile phone, for example, may be practical and immediate. However, it is hardly creative, as the phone is far from the content world of a change of wheel. Also, this solution is not always practical, as it will not help those not owning a mobile phone. The idea of catching a ride is similar to that of the mobile phone: if the road is deserted, it is not relevant; and if cars were driving past, the idea would arise without the need for creative thinking. Note the reverse relationship between the distance of the resource from the content world of the problem and the level of creativity of the solution: the more distant the resource, the less creative it will be considered.

Although both conventional ideas mentioned might help the driver, most of them do not solve *the problem we have raised*: they may aid the driver in continuing on her way, but the wheel has not been changed. Let us examine other ideas that try to tackle the problem directly. The idea of using oil from the engine or the brakes to oil the bolts so that they will come "unstuck" was ranked high by the group members, and received the proximity rating "medium." Note that such an idea already belongs to the content world of the car.

Table 6.1 Ideas for solving the problem of changing a wheel

Idea	Proximity	Creativity
Using mobile phone to get help	Far	Low
Catching a ride in order to get home	Far	Low
Using pipe to lengthen crossbar (increase leverage)	Medium	Medium
Using oil from engine or brakes to clean rust	Medium	Medium
Backing up with crossbar on bolt to increase force	Medium	Medium
Attaching the jack to the crossbar to help turn it	Near	High
Raising the car, placing stone between crossbar and ground, then lowering bar over it to add force	Near	High

If we accept the principle *that the nearer we draw, in the search for a solution, to the core of the problem, the higher the "grading" in creativity*, then elements from the content world of changing a wheel should be good candidates for a creative solution. Let us look at the solution of using the jack: this tool comes from the world of changing a wheel, and is an interesting candidate for a solution. The idea suggests that we use the jack (an existing resource) to exert force to turn the crossbar. The jack can exert much force (it is designed to lift a car), and we may therefore be able to break loose the rusty bolts. Once the bolts have been loosened the jack can be used in its original role of lifting the car for changing the wheel.

We may deduce from Table 6.1 and from the above discussion that the most creative solutions are those found "at hand," i.e., near to the content world of the problem. This is the reason why many times, when a creative idea is presented to us, we mumble, "Why didn't I think of that before?" Again, we find that creativity is about an intelligent search among a limited list of possibilities. In Attribute Dependency we searched the list for two suitable (unrelated) variables; in Replacement, we will look among the components (in the immediate environment) for one that may fulfill the required role and aid us in reaching a solution.

The principle that connects the proximity of the solution resource to the core of the problem and perceived creativity was discovered by Maimon and Horowitz [1]. They reported two sufficient conditions for inventive solutions in the context of technological problem-solving, one of which is the "close world condition." According to the close world condition, only certain predefined resources that are present in the proximity defined by a system may be considered in the search for an inventive solution. In our framework the close world is not dichotomous, rather, creative perception can be viewed as smooth effect. The closer the resource to the product, the more creative it is perceived.

Here is another example to clarify our point. Imagine that an open field was set on fire. If we had fire engines, fire extinguishers, blankets or sand-pails, we would use them to extinguish the fire. These are all conventional solutions and, being so efficient and familiar, they are not considered creative. In certain cases, when no other means is available for extinguishing the fire, we might use an existing resource – the fire itself. Farmers sometimes burn a thorn-field, and the scorched earth serves as a partition stopping the spread of the fire. This is a more elegant and creative idea, because it uses a resource in close proximity to the problem: it is in fact the problem itself.

> The Replacement Template is based on the replacement of a resource or component existing in the system or in its **immediate** environment in order to fulfill a necessary role.

Replacement Template in nature

An interesting story illustrating the Replacement Template is that of the hermit crab. This tiny crab is not equipped with a hard shell and it is therefore prone to attack from various predators, who may easily enjoy its soft meat. In order to solve this problem, nature "uses" the Replacement Template. This crab, living by the sea, houses itself in the empty shells of oysters or snails that are plentiful all around it. As it grows, it extracts itself from the shell and searches for a larger one. Thus, the hermit crab uses resources available in its natural environment to serve a role crucial for its survival.

Implementation of the Replacement Template

In order to define the exact thought process that underlies the Replacement Template, let us return to the pizza example. Bear in mind how Attribute Dependency helped to identify an interesting idea that could present the competitors with a dilemma: the price of the pizza would be dependent on its temperature. One of the difficulties stemmed from a physical problem: the pizza would lose heat in spite of the insulated cardboard box if delivery took more than 30 minutes.

From a logical standpoint, this problem may be solved in one of two ways:
1. Better insulation.
2. Adding heat during delivery.

The first solution would represent "more of the same." This idea does not suggest any radical change, but a gradual improvement in the existing system.

The second solution clearly points to the basic lack of a resource which, if found, would be a more creative solution. From the minute the pizza is removed from the oven in the pizzeria, it loses heat continually until it arrives at the client's house. One of the suggested solutions is to use the engine of the motor scooter (used for the delivery) to heat the pizza on the way. This solution usually arouses antagonism at first, because one immediate reaction to the idea might be that the exhaust fumes would contaminate the pizza while heating it. Such recoil is not a reason to discard the idea. One must first identify the benefits and positive aspects of the idea and, if they are worth it, ways must be searched for to solve the problems involved. The benefit in using the heat of the engine is considerable, and may be even extrapolated beyond keeping the pizza hot: theoretically, one may even bake the pizza on the way to the client, thus saving valuable time.

There is a great advantage to this idea. Shifting focus from the time of the pizza's arrival to its temperature will pose a serious dilemma to the competing chain. The next announcement might be that our delivery time is 10 minutes less than our competitor's: if we could cook the pizza in transit we might be able to guarantee delivery in under 20 minutes. Such dramatic moves greatly increase the benefits to the customer. It is hard to foresee whether they will enhance the establishment of the new company in the market of pizza delivery, but they certainly afford food for thought in consolidating its strategy.

Now that we have found the benefits of the new idea, we may proceed to the technical planning of the necessary source of heat. We will turn to our heat engineers asking them to search for possibilities of building a device to heat the pizza without bringing it in contact with the fumes of the scooter. In fact, this engineering problem is not too difficult to solve. The decision whether to go ahead with the development of the idea will have to take into account market considerations such as the image of a pizza baked on a scooter engine. If we find out that customers are opposed to the idea, we will have to look for other possibilities of heating the pizza during delivery (e.g., using the electrical system).

Replacement as a Template of Innovation

In the pizza case the Replacement was utilized to solve an actual problem. Normally we would like to use Templates as idea generators. Now we will demonstrate a way to reverse the order of thinking: arrive at original ideas without previous knowledge of the problems they may solve, on the basis of the

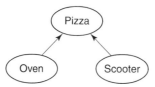

Figure 6.4   Configuration of pizza delivery system.

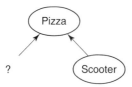

Figure 6.5   First stage in implementing Replacement Template on pizza delivery system.

principle of Function Follows Form (presented earlier, in the context of the Attribute Dependency Template), the benefits will be discovered at the second stage. We will again use the pizza example. Let us assume that we are not aware of a certain problem in our product, and we wish to think of ways to improve it.

We may partially describe the delivery system by a sketch of its components and the links between them (see Figure 6.4). Such sketches contain internal information (about the product or service) embedded in it during its development, like the information contained in a matrix. The ellipses in the figure contain components of the product, and the arrows show the link between them.

In order to understand the structured thought process that underlies the Replacement Template, let us imagine a virtual problem in which the oven is out of order and it is impossible to bake pizzas in it. If we wish to prevent closing down the pizzeria altogether, the natural thought process is to search for alternative sources of heat. The significance of this decision is that although we have relinquished the oven as a component (a means for baking), we did not relinquish its function (baking pizzas). We shall therefore remove the component but keep the arrow expressing the role of the removed component (Figure 6.5).

It would seem that the number of alternatives to an oven (as a source of heat) is great, but we can actually narrow down the search for such an alternative. The new component should include two basic characteristics:

1. **It must be local.** The more available it is (and nearer the content world of the problem) the more creative it will be deemed.

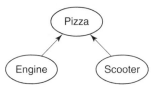

Figure 6.6   Replacement of role of oven by scooter component.

2. **It must carry out the required function**. Therefore, its characteristics
   should be similar to those of the removed component (we shall return to
   this requirement later).

The engine of the motor scooter (and also its electrical system) fulfills both of
these conditions. It belongs to the content world of pizza delivery and it produces heat as part of its regular function. Figure 6.6 describes the new situation in which the engine fills the role of the oven.

We have learned in this process that it is possible to shorten the time of
delivery even without encountering an actual problem, through a hypothetical thought process. By using the Replacement Template on a sketch of the
system, a new system was designed. Thus, it is possible to use Replacement as
a model for generating innovative changes in an existing product. The question the marketing experts must ask themselves during this process is similar
to that asked in Attribute Dependency: How can the new system be used, and
what new benefits does it provide?

In order to implement the Replacement Template in a systematic and structured way, we should use the possibilities inherent in it by operating the
Template in reverse order, as we did in Attribute Dependency. We shall first
operate the Replacement (almost technically) on some system, then scan the
space of possible benefits and check whether we have solved the problem, or
whether we have created a response to a new need. Let us first redefine the following terms:

1. **Components:** Self-standing parts or sub-systems (static objects). The components of a car are the wheels, engine, seats, etc. Note the difference between
   a component and a variable: the wheel of a car is a component, while the air
   pressure in it is a variable. Other examples may be found in Table 6.2.
2. **Internal component:** A component over whose variables the manufacturer
   or the service deliverer has full control. The car seat, for example, is an
   internal component because its softness, color and size are determined by
   the manufacturer. (See Table 6.3 for some examples.)
3. **External component:** A component outside the control of the manufacturer, and in direct contact (in time or place) with the product. For

Table 6.2 Examples of components and variables

Component	Variable
Chair legs	Height of chair legs
Automobile fuel	Fuel quantity, fuel temperature, type of fuel
Cupboard handles	Type of material of handle
Writing in advertisement	Size or color of lettering
Picture in newspaper	Size or color
Milk in coffee	Amount of milk, fat content
Moisture in ointment	Amount of moisture
Filling of pillow	Type of filling

Table 6.3 Internal and external components

Product	Internal component	External component
Mattress	Springs	Material of sheet, person lying on it
Chair	Chair legs	Upholstery of seat, height of chair legs, floor
Suntan lotion	Radiation filter, moisturizing substance	Skin of user, radiation of sun
Electric boiler	Heating element inside boiler, insulation material	Water, sun
Camera	Flash, focusing lens	Outside light, subject of photo

example, an automobile manufacturer has no control over the road. However, the latter has a direct connection to the car: it comes in physical contact with the product at a certain time and in a certain place. (See Table 6.3 for some examples.)

4. **Link** between two components exists in the following circumstances:

   (a) One of the components controls the variable of the other.

   (b) This control is directed and required by the design of the manufacturer or deliverer of service.

   A controlling component is one that determines the value of a variable in another component (the controlled component). When the link between the two is removed, the value of the variable in the controlled component will change. Such is the link between two components of a light bulb: the electric current and the incandescent wire. The electric current (the controlling component) determines the values of the variables "color" and "temperature" of the incandescent wire (the controlled component).

5. **Configuration of product:** The complete set of all the links of a product is

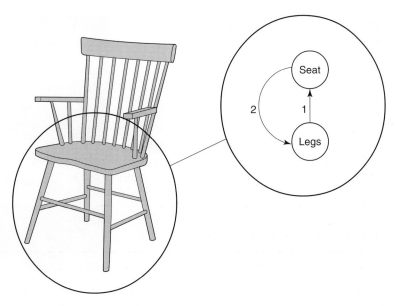

Figure 6.7   A link between legs and seat of chair.

defined as its configuration. It is easy to draw a sketch of a configuration in the form we have seen for the pizza. Naturally, all configurations (including those in this book) are partial. We will see later that there is no need to include all the potential external components of the product in order to operate the Replacement Template upon it.

Case study 1 – a chair

In order to explain the way the Replacement Template is implemented, let us examine a basic product, a chair, and "predict" with its help an already existing product. We will first construct the product configuration of a typical chair in the following way: We will denote the components by circles and the links by arrows, with the direction of the arrow determining the type of relationship between controlling and controlled components (direction of arrow – from controlling to controlled).

A partial (yet representative) list of the internal components of a chair includes legs, seat and backrest. The external components include floor, wall and user. Let us first examine two internal components: legs and seat (see Figure 6.7). The relationship between the two components obeys both rules of link defined above: the legs control the seat, as they determine its height.

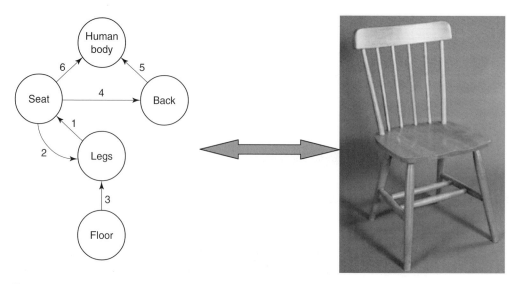

Figure 6.8　Product configuration of a chair.

This control is determined by the manufacturer, thus it is a legitimate link (denoted "1"). In addition, the seat (in many types of chairs) supports the legs and holds them in place, and this is an additional link (denoted "2"). Note that the legs act as a controlling component in link "1", and as a controlled component in link "2".

In order to construct the product configuration of a chair, we must collect a complete set of the existing components and the links between them. Since we include in the configuration the basic links of the components of the product, this is not a complicated undertaking. The wall, for example, has no planned link with a standard chair, therefore this external component will not be found in the configuration.

Figure 6.8 represents the full configuration of a chair (for typical chairs). Additional functional links of the chair configuration are detailed as follows:

3. The floor supports the chair legs and stabilizes it.
4. The seat holds the backrest and positions it in space.
5. The backrest supports the back (for comfort).
6. The seat supports the buttocks (at the desired height).

This chair configuration includes all the information needed for operating the Replacement Template. In order to locate the replacement resources, we will first remove an intrinsic component from the configuration, not yet thinking about benefits or aims. The definition of an intrinsic component is quite simple: it is a component without which the product loses most of its

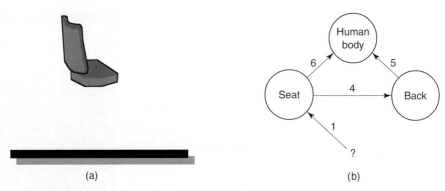

(a)   (b)

Figure 6.9   The first stage in operating the Replacement Template on a chair. (a) Physical model. (b) Intermediate product configuration of chair.

advantages and becomes useless. The chair has two essential components: seat and legs. Let us remove one of them from the chair configuration without denying its function as a controlling component, just as we removed the oven in the former example without forgoing its function (link) as the baker of pizza.

> Operation of the Replacement Template includes the exclusion of an intrinsic component together with the links in which it functions as a controlled component, without removing the links in which it functions as a controlling component.

Removal of the legs (keeping the dangling link) from the chair will result in a chair "floating in air at the desired height" (see Figure 6.9). This is an absurd thought construct, but it will serve us temporarily until we operate the Replacement Template to the full, after which the chair will of course not float in the air. The function of the legs – fixing the seat at a certain height – must be fulfilled by another environmental element. As at this (thought) stage the chair "floats," the floor has no contact with it and therefore has no functional link to the configuration components. Note that in the product configuration, the floor functions as a controlling component and the legs as the controlled component. Therefore, when we remove the legs from the configuration, the floor is removed with them. However, removal of the legs does not remove the seat, as there is the link we termed "1" in which the legs were the controlling component over the seat. Link "2," in which the legs functioned as the component controlled by the seat, is also removed from the configuration. The result of this operation is shown in Figure 6.9.

In order to operate the Replacement Template successfully, we shall try to locate a component to replace the missing one. Where shall we search, and how shall we identify it? The missing component will be an external one, in contact with the product, **visually or functionally similar to the missing component**; and the site of search will not exceed the bounds of the content world of the product. The reasons for this are:

1. When we remove a component from the product configuration, we leave "orphaned" links, in which the removed component functioned as a controlling component. We need a replacing component to assume these links, therefore it must be physically or even functionally similar to the one removed.

2. Searching for a solution in the immediate environment of the product is considered a more creative strategy. In the flat tire example, we illustrated a principle in which the nearer the solution is to the content world of the problem, the more creative it will be considered. This claim is supported by empirical studies that examined cognitive concepts of creativity. These studies found that the nearer the proposed solution, the more brilliant and creative it will be considered.

3. Components in the immediate environment of the product are more easily available, and the process of locating them will be quicker and more focused.

We must therefore find among the components present in the immediate surroundings of the chair the component most similar to it, to place in the dangling link. The possibilities are quite limited: wall, table, carpet, user, floor – all external components that come in direct contact with the chair and are not under the control of the manufacturer. The table stands out in this list in its similarity to the chair, both in its engineering concept (it, too, has legs) and in the functions it fulfills (support of objects at a certain height).

Replacement can create a change in the internal components of the chair in such a way that the table (an external component) will replace the missing component (legs). The new product configuration and its description in the physical world are represented in Figure 6.10.

This new configuration of a legless chair will be of no value unless some practical use can be found for it for a defined target population. We must remember that it is plausible that we will find use for it. This statement is based on the claim that products obeying "Creative Templates" have good chances of implementation, as well as on the principle of Function Follows Form – that a thought process leading from configuration to benefit is effective. It is fairly easy to recognize that a target population for this chair may be parents of babies, who might be interested in the following benefits:

1. Whatever the dimensions of the table (high or low) the baby will sit at an appropriate height in relation to the table.

2. The chair will be easy to carry around, because it does not have a cumber-

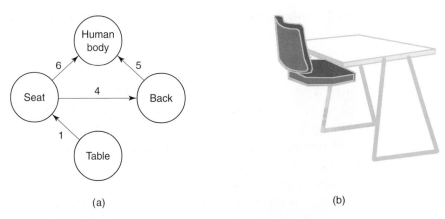

(a)   (b)

Figure 6.10   Replacement of the function of the legs by a table as component. (a) New chair configuration. (b) Physical model of new chair.

some component. (The legs of a high-chair are longer and heavier than those of a regular chair.)

3. The floor under the chair can be cleaned more easily than under a normal high-chair.

Before we proceed further we would like to draw attention to two important concepts of the Replacement Template.

(1) Why must an intrinsic component be removed?

The thought of removing an essential component is usually accompanied by feelings of anxiety or insecurity, as it is the very heart of the product and it seems that its elimination will cause the collapse of the whole product concept. The fear of removing such a component is emotional only and is not logically justified, for the following reasons:

1. The hypothetical exclusion of a component does no physical harm to the product – it is not really damaged.

2. Statistics points to a great wealth of ideas connected with such exclusion, therefore a thought effort in this direction is justified.

3. The function of the removed component is not damaged, so that the product will not "collapse." The function will be replaced by an alternative component (according to the Replacement Template), therefore there is no danger of losing the product. Even when we removed the legs from the chair, it remained at the appropriate height. If no other component can be found to fulfill the function, the component will be returned to its rightful place.

(2) The importance of the locality principle when matching the external component

Consider the hypothetical product in Figure 6.11 (this concept was spread on the Internet as part of new product jokes). The matching between the

Figure 6.11   A forbidden Replacement.

shaving machine and the cellular phone technically may be considered as another manifestation of a Replacement Template. A closer examination reveals that there is a remote association between the removed part of the phone and the shaving machine. Indeed this absurd product loses its meaning exactly because the huge distance between the removed component and the imported one. Creative ideas in the context of new products have to also meet market requirement and the rules of similarity and vicinity of the imported external component keep this requirement.

Case study 2 – a scanner

The chair example may give the impression that the Replacement Template is suitable for lo-tech cases. The next example is a real field application in the context of high-tech products [see 3].

Replacement in a scanner

Figure 6.12 below depicts a basic part of a scanning system, which relates to a low-priced scanner. A mirror directs light beams that carry the information of a scanned picture towards a CCD element. The CCD (charge-coupled

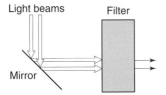

Figure 6.12   A partial trajectory of light beams in a scanner.

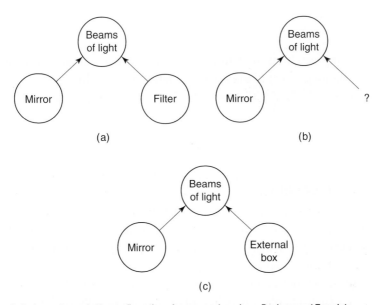

Figure 6.13   Inducing a change in the configuration of a scanner by using a Replacement Template.

device) transforms the optical data to electronic signals and, in order to protect the relatively sensitive CCD element, a filter absorbs an essential amount of the energy of the beams.

As part of developing a "low-price" new model, a trained engineer was asked to suggest a way to reduce the manufacturing costs. Figure 6.13a–c depicts the three steps in the implementation of the Replacement Template. In Figure 6.13a a (partial) configuration is presented; in Figure 6.13b one of the intrinsic components (the filter, which was expensive relative to the mirror) is eliminated. According to this Template, its function was not removed. The function of absorbing the excess radiation was then assigned to the body of the scanner. The resultant new configuration is displayed in Figure 6.13c.

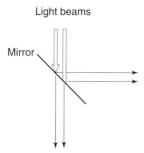

Figure 6.14   The new optical system.

The implementation included an altered mirror which reflects only part of the radiation, allowing the excess radiation to pass through and be absorbed in the scanner's body (Figure 6.14). The reduction in costs due to this change was estimated as 5 percent of the manufacturing cost. The new scanner was introduced to the market during the winter of 1997.

Case study 3 – butter patties

Templates may be used also when we want to introduce simple yet original innovations in systems, or even when we want to solve problems that are well defined [see for details 1, 2, 3, 4].

Let us portray the notion of Creativity Templates by an example of a field case, in which a Replacement Template was successfully applied. In the process of manufacturing butter patties the patties that do not meet specifications are recycled by placing them in a jacketed metal vat and melting them by steam (Figure 6.15 below). This conventional process is slow, due to the low heat conductivity of the butter and the airspace between the patties, that acts as a heat insulator. Increasing the efficiency of the steam requires further development of the current system (e.g., increasing the temperature of the steam, developing a better contact area between the tubes and the butter).

The Replacement Template – a step-by-step presentation

(1) Draw the configuration of the system

The set of relevant components and their (designed) functions (defined by links) compose the configuration of the system that captures all the necessary information for the Template. A partial configuration of this system is

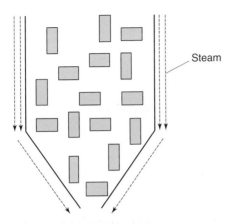

Figure 6.15   Butter patties in the vat before melting.

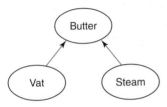

Figure 6.16   A partial configuration of the melting system.

presented in Figure 6.16. The link between the vat and the butter consists of the vat's function, which contains the melted butter. The function underlying the link between the steam and the butter is producing the heat and transferring it to the butter for the purpose of melting.

(2) Split an intrinsic link and exclude one intrinsic component

At this step *an intrinsic component* is eliminated from the configuration while preserving its associated intrinsic function. In the resulting intermediate configuration this intrinsic function is not carried by any component and is termed the *unsaturated intrinsic function.*

In this example both the steam and the vat are intrinsic to the system. However, the slowness and lack of efficiency of the melting process is caused by the heat transfer system (the steam) rather than the system which is designed to contain the melted butter (the vat). The result of applying this step on the steam is shown in Figure 6.17.

(3) Assigning the unsaturated function to a suitable component

A suitable component is defined as a readily available component (in the system or in its immediate environment) whose associated function or

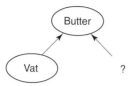

Figure 6.17   A dangling link after eliminating the steam while preserving its function.

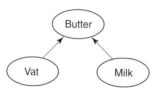

Figure 6.18   Assigning the function of heating to the milk.

intrinsic attributes are similar to the removed component's attributes or function (i.e., the unsaturated function).

Since the list of readily available components is limited, one can proceed to an exhaustive search for suitable components. Normally, this search should be efficient due to the constraining requirement of similarity. In our example, consider the hot milk contained in the first station of the butter manufacturing process. The similarity between the milk and the steam lies in the fact that both are hot fluids. In fact, this is the only component in the environment that is similar to the (removed) steam. The result of assigning the heating function to the milk is a new configuration in which its intrinsic function is being saturated (Figure 6.18).

According to the new idea the hot milk (from which the butter is produced) will be exploited to melt the recycled patties. Instead of using outer tubes with hot steam, the hot milk can flow inside the vat and directly melt the butter. The resulting mixture of butter and milk then flows back to the beginning of the butter manufacturing process without further changes.

The complexity of the above technological system and its cost functions (e.g., maintenance) have been dramatically reduced (the steam system was removed from the process). Yet, the process of recycling patties was shortened due to the increase in the efficiency of the heat transfer to the butter (by introducing a direct contact between the new melting system and the butter), thus time and energy costs are reduced. Technology changes like this, although transparent to butter consumers, may increase the flexibility of the manufacturer to offer different service or price, or simply improve the cost-benefit function.

When is exclusion appropriate?

If we look again at the examples of the lighthouse and the antenna pole, we shall discern that what transpired in them did not involve the exclusion of a component. From a practical point of view, we may say the exclusion is not vital to Replacement. Only in cases where the thought process involves searching for a new product without facing a defined problem will we carry out exclusion, in order to generate an artificial problem from which a new idea may emerge. We did so in the example of melting the butter, and this led to a classical case of exclusion, in which steam was no longer part of the configuration of the vat.

However, if we define a more coherent theoretical structure, we may claim that exclusion also exists in cases in which Replacement is applied to solve an existing problem that does not appear to respond to the pattern of this Template. In order to do this, we will treat exclusion slightly differently. Almost any problem may be solved by means of a non-creative or non-applicable idea (a distant cousin who will change inscriptions on the lighthouse, or concrete on the antenna). This imaginary resource is "a component signifying the function:" after its exclusion, the function will be fulfilled by Replacement. By the apparent addition of a component representing a routine idea and its later elimination, we can operate the Replacement Template and reach a more creative idea.

Case study 4 – Nike-Air® ads

In Chapter 11 we will discuss the validation of the Templates paradigm and techniques. Meanwhile we would like to show how the same Template (with small modifications) appears in a remote field of expertise in which creativity plays a central role: advertising. This claim is not very surprising in light of the fact that Replacement structures were identified both in new products and technological problem solving contexts. If correct, the universality nature of this Template may impinge on how we perceive creativity and how we should understand it.

To illustrate the operation of the Replacement Template in the realm of advertising, consider the advertisement of Nike-Air® shoes (see Figure 6.19). The shoe has a trait of "cushioning and absorbing the shocks" caused by various sport activities. The visual component in this ad shows a group of firemen holding a shoe serving as a life net for fire victims escaping from a

The Air Essential. Something soft between you and the pavement.
INTRODUCING TWO NEW WALKING SHOES FROM NIKE. WITH NIKE-
AIR® CUSHIONING IN THE HEEL, THEY'RE VERY SAFE PLACES TO LAND.

The Air Healthwalker Plus

Figure 6.19   Nike-Air® (Wieden and Kennedy, 1992, *The One Show* [5]).

burning building. The Replacement Template is obtained when a product P
(e.g., sneaker) or one of its aspects A (e.g., shape), *replaces* the corresponding
aspect of the symbol S (e.g., fireman's life net) in a situation where its trait (T)
("cushioning and absorbing shocks") is crucial (saving persons from being
crushed). The aspect A substitution can be represented by a link between a P
and a S.

The general scheme and the implementation in the example of Nike-Air®
are given in Figure 6.20. Figure 6.21 illustrates how a sequence of the four elementary operators – *split, exclude, include* and *link* – generate a *linking operator.* The exclusion operator removes an attribute (or a component, e.g., the
firemen's life net is excluded from a rescue situation) after being split from its
links. The inclusion operator introduces a new element, S, to the environment
under consideration (e.g., if the environment is a rescue from a burning building or a suicide situation, a Nike-Air® shoe is an included component). The
linking operator substitutes the excluded component by another (e.g., using a
shoe as a net).

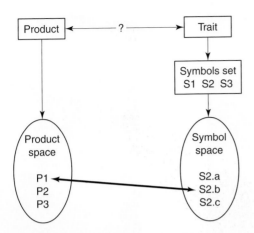

Figure 6.20   A general scheme of the Replacement Template.

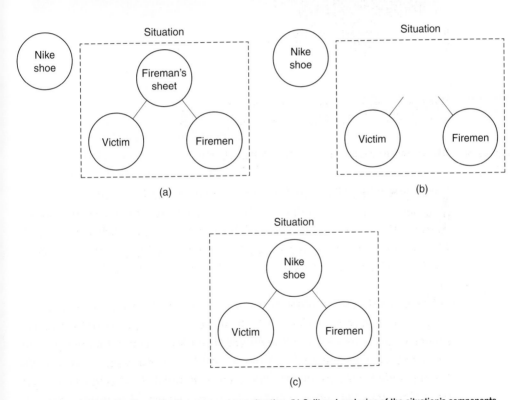

Figure 6.21   (a) Inclusion of a product's component to a situation. (b) Split and exclusion of the situation's components. (c) Linking.

Figure 6.22   Bally shoe (Wieden and Kennedy, 1992, *The One Show* [5]).

Case study 5: Bally shoe ads

The recurring incidence of the Replacement Template may be exemplified by the Bally shoe ads in Figure 6.22. The advertisements are intended to associate the shoe with the sense of freedom for the foot by replacing the contour of an island or clouds (symbols of freedom) by the shape of a foot. Although at first glance it may appear remote, the ad's creative concept for Bally shoes, shown in the figure, can be interpreted as having the same fundamental scheme as the Nike-Air® sneakers ads.

Replacement vs. Attribute Dependency

In operating the Replacement Template the chance of locating a broad store of ideas is small, compared to what often happens in using the forecasting matrix. Sometimes, use of the Replacement Template does not yield a single idea. However, ideas of Replacement usually require little investment, and their average originality is greater than that of ideas born through the use of a prediction matrix. Replacement is similar to risk investment: possibly, after five hours of thinking not a single idea will be generated; and possibly, two hours of thinking will yield a much better idea than many others that may immediately pop up in using a prediction matrix. An indication of cases in which it is worthwhile to prefer Replacement to Attribute Dependency is the presence of a saturated matrix. If no space is available for Attribute Dependency, it seems worthwhile to start excluding components.

Operational prescription

Operational prescription

1. Make a list of internal components (over which the manufacturer has control).
2. Make a list of external components (over which the manufacturer has no control).
3. Construct a product configuration, mapping all the desired control links between the listed components.
4. Locate the essential components, mark them and write their function in all the links.
5. Choose one essential component, exclude it from the configuration, but leave the function it had fulfilled.
6. Scan the nearby environmental components and make a list of components having characteristics or functions similar to those of the excluded component.
7. Connect each component from the new list to the function missing a component. Describe a physical model of the new configuration.
8. Search for a new market benefit derived from the replacement you have created. Remember that the principle of Function Follows Form supports you. Do not overlook any new market shares that may result from the added dimension.
9. According to your success, decide whether to continue linking other environmental components to the missing link, or exclude other essential components.

REFERENCES

1. Maimon, O. and Horowitz, R. (1999) "Sufficient condition for inventive ideas in engineering," *IEEE Transactions, Man and Cybernetics*, **29** (3), 349–361.
2. Maimon, O. and Horowitz, R. (1997) "Creative design methodology and SIT method," *Proceedings of DETC'97: 1997 ASME Design Engineering Technical Conference* September 14–17 1997, Sacramento, California.
3. Goldenberg, J., Mazursky, D. and Solomon, S. (1999) "Creativity Templates: towards identifying the fundamental schemes of quality advertisements," *Marketing Science*, **18**, 333–351.
4. Sickafus, E. (1996) "Structural inventive thinking: a conceptual approach to real-world problems," *The Industrial Physicist*, March, American Institute of Physics.
5. *The One Show* (1992) Vol. 14. Switzerland: Rotovision, S.A.

7 The Displacement Template

What is the Displacement Template?

The Displacement Template states that a component may be eliminated from the configuration of a system **along with its functions,** and thus a new product will be created, targeted to a new market. This stands in contrast to the Replacement Template in which the function (represented by a link) of the removed (intrinsic) component was kept in the configuration. Before presenting the implementation and principles in detail let us illustrate this Template with some examples.

In the early 1970s, after a lengthy marketing campaign, the efforts to introduce food products based on a powder for instant home preparation (among them, of course, instant coffee and various soup mixes) into the US market finally succeeded. Among the products introduced in the market was a cake mix for home baking. The idea was that a customer buys the special powder, mixes it with water, pours it into a baking pan and puts it in the oven. Expectations for success were great, both because the product seemed to capture a large market share, and because all the tests gave its taste a high grade. To the manufacturer's great surprise, the marketing efforts failed.

Extensive market research discovered that although US customers wished to save time in food preparation, they were interested in adding their personal touch to the cakes they bake, to give the cake a special "home-made" taste. In order to meet this need, it was suggested that a bit of the product's final form be "eliminated". After considerable consideration it was decided to remove the eggs from the mix, and instruct the amateur bakers to add them themselves thus alleviating the objection to a factory-made cake. Now that the customer had to expend the effort, sales of the cake mix rose markedly.

Another example of the success of the Displacement Template is the vacuum package, which is based on a simple principle: removing the air from the packed product. Thus, the freshness of food products, such as meat and natural juices, is preserved. Keeping the packed products from coming in contact with the air enables a longer shelf life, and makes it easier to store and use them.

Another successful product developed by removing a component from an existing brand is Motorola's cellular telephone, known as Mango®. In this device, the possibility of an outgoing call was removed, markedly reducing both the price of the instrument and the telephone bills, and broadening the target population of cellular communication. For example, parents' concern about excessive use of cellular phones by their children was eliminated enabling the children's segment to be targeted (a highly important group for companies wishing to acquire long-term loyal consumers).

A fruitful idea: the Mango®

Israel is ranked very high in the per capita adoption rate of cellular phone lines. The accelerated development toward a saturated communications market requires innovations on the market, and most of them take the form of sales projects by the main operators.

The story of the Mango® cellular phone is an example of a technologically simple and brilliant success in the market which, it seemed, was quite a surprise. The vice-president for marketing in the firm came upon this innovative idea when searching for an appropriate response to the low prices offered by a competing firm. He identified a sizable market segment that was not interested in registering, identifying itself and receiving phone bills from yet another communications company. The needs of this market segment were limited, as must be the instrument offered to them: a simple and cheap instrument sold in the supermarket, through which one may receive phone calls free of charge, while communication from it is limited. This concept could be implemented after the billing regulations in Israel changed and the billing of an incoming call was changed to the person who calls. The result was a great success: in less than one year, more than 5% of the population purchased Mango® instruments. The international edition of *Advertising Age* in 1995 chose the marketing move of Mango as one of the 12 most brilliant moves worldwide for that year.

The Replacement Template introduced us to the idea of excluding internal components, without elimination of their functions. In contrast, the Displacement Template, demonstrated in the above examples, is characterized by the exclusion of a component **without** replacing its function by another component.

The Displacement Template refers to the removal of an intrinsic component from the configuration including its functions, in a way that causes a qualitative change in the configuration.

Let us elaborate on this difference. In the example of the chair analyzed in Chapter 6 (Figures 6.7, 6.8, 6.9) we presented the removal of the legs from the configuration of a chair while retaining their function, in the spirit of the

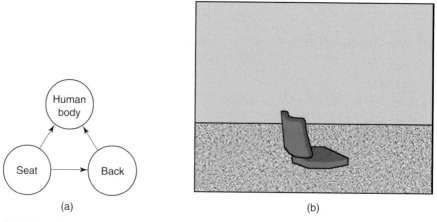

(a)   (b)

Figure 7.1   The first stage in the operation of the Displacement Template upon a chair: removal of the legs and their function. (a) The resulting configuration. (b) The physical model.

Replacement Template. In the Displacement Template, we exclude completely all the functions of the missing component. Figure 7.1a describes the configuration of a chair after removal of its legs and elimination of their function. The physical model of this configuration is shown in Figure 7.1b. The chair does not "hang in mid-air" as it did in the Replacement Template; the function of the legs (support of the seat at a certain height) is not fulfilled and the chair seat is therefore at ground level.

The exclusion of an essential component from a product causes a reduction in its performance, and may be quite detrimental. Actually, this legless chair is a problematic product. Action similar to that carried out for the cone-shaped glass (Chapter 5) is now in order: we must search for a population group that may develop a demand for this exceptional product. This line of thinking, previously referred to as Function Follows Form, requires suspension of an intuitive judgment about the creative ideas. No idea will be rejected before it is carefully examined for all its inherent possibilities.

Readers are probably familiar with such a legless chair: It is used to solve the problem of using a normal chair at the seashore, when the legs dig into the sand and cause the chair to be unstable and uncomfortable to sit on. We encountered a similar problem when we discussed Attribute Dependency in the glass, when we suggested a "bottomless" cone-shaped glass. The solution here is similar: since we are used to sitting on the sand, we can dispose of the legs and use the chair without them. Such a chair has an additional advantage: it is easier to carry.

Another blockbuster product, which shares the Displacement structure, is

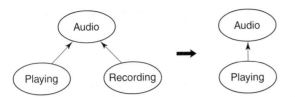

Figure 7.2   The changes in the configuration of the Sony Walkman® due to the Displacement Template.

the Sony Walkman® [for a detailed description see 1]. In this case too, the invention was not initially intended. Indeed, even during the first stages following its introduction, the marketers did not envision the success potential of this product. The earlier monophonic Pressman® which had a recording device failed and was abandoned. Attempts to make it smaller failed too, because the recording system did not fit into its small size. Rather than invest any more effort, it was put aside and used by the company's engineers for their own entertainment. Only after the integration of this concept with that of a light headphone, was the Walkman® concept defined. Incidentally, market research failed to predict the tremendous success of the Walkman® concept.

Figure 7.2 illustrates the basic structure of the Displacement Template in the Walkman® case. It is interesting to note that this invention also makes use of the Replacement Template: the cord of the earphones is used as the radio antenna.

Displacement is not unbundling

It is very important to differentiate between the Displacement Template and a marketing approach called "unbundling." The unbundling approach is one by which one may **quantitatively** diminish the components of a product or deliver a service in a way that detracts from its efficiency or quality, and charge a lower price for it. This approach enables the enlargement of the target population using the product by including a lower-income or price-sensitive segment. An example of such unbundling is a lower standard tour package (using cheaper hotels and charter flights) offered at a low price; or "do-it-yourself" furniture (entailing a lower level of service and therefore a lower price). Note that no new benefit is created by unbundling, except the lower price. In Displacement, on the other hand, there is a new benefit connected with the characteristics of the product (sometimes the price does not vary

Table 7.1 Displacement versus "unbundling"

Product	Displacement	"Unbundling"	Benefits
Monocycle	×		A new sport
Insurance plan requiring lower fees, but offering less coverage		×	
Walkman® (Sony's first model of Walkman®)	× ×		A mobile and compact radio
Suntan lotion – less protection to give a better suntan in winter	× ×		Suitable for winter and transition seasons
Soapless soap	×		Healthier for skin
Suntan lotion with lower protection coefficient at lower price		×	
Calendar without days, for multi-annual use	×		Economical usage
Decaffeinated coffee	×		Healthier, not a stimulant
Fingerless gloves	×		Allow free movement and sensitive touch of fingers

from that of the original product). Table 7.1 provides additional examples for differentiating between the two cases.

Implementation of the Displacement Template

Another example of the Displacement Template is shown in Figure 7.3, describing a partial configuration of a regular television set having the following connections: the box and the remote control determine the position of the various parts and the timing of their function. The receiver, loudspeaker and screen translate the television broadcasts to motion picture.

Intrinsic components of this configuration are the receiver and the screen. Exclusion of the screen from the configuration along with its function will bring about the product configuration shown in Figure 7.4a. Figure 7.4b represents a physical model of a television set with no screen. This set can be very small, because the screen takes up most of the space.

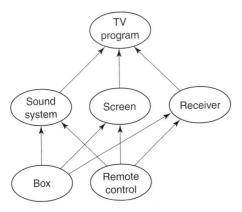

Figure 7.3   Product configuration of a television set.

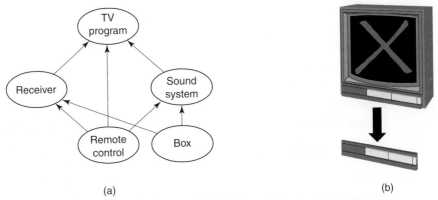

(a)   (b)

Figure 7.4   Removal of screen from television set. (a) Product configuration of TV without screen. (b) Physical model of configuration.

Is a TV set that displays no picture actually a radio set? Not necessarily: This appliance does function exactly like a radio receiver – the wavelength it receives and the content of the programs it receives belong to the TV world (and not to the radio world – not even radio that occasionally broadcasts television programs). In fact, even its electrical components are different than the radio components. The marketing question defined now is who needs an apparatus that can receive TV programs without showing them visually. Surprisingly enough, a market niche was found for this product: it serves drivers on their way home who do not want to miss their favorite TV programs.

We have already stated that many creative ideas are available "at hand." Similarly to the Replacement Template, the Displacement Template can also reveal creative and simple solutions to problems and needs that the market or the producer are not yet aware of. Use of this Template simplifies an existing product by excluding components, thereby gaining two distinct advantages: (1) R&D efforts are less than those in the addition of new dimensions to a product; (2) customers are already familiar with the product in its sophisticated version, and there is therefore no need to present a new and unfamiliar product to them.

Observations on the Displacement Template

Caution with Displacement: a Displacement without a real exclusion pitfall

At the beginning of the 1970s "smokeless cigarettes" appeared on the market. This was a novel and revolutionary product: A cigarette that looks like any other and contains the usual amount of nicotine with an excellent taste and aroma, but that does not emit smoke. Great resources were invested in research and development of this product, and advanced technology was employed. The product was presented to the market as carrying a new message – the end of discomfort for the non-smokers – and was accompanied by an extensive advertising campaign.

Note that this story does not represent the Displacement Template. The smoke component was not excluded from the product, but was substituted by another component, with a substantial investment in R&D. Although at first glance the change appeared to be that of Displacement, it was in fact a completely different product.

Although it seemed like a brilliant and far-reaching idea, this cigarette was in fact completely rejected by the market [2]. Smokers were not convinced of the benefits of switching to a smokeless cigarette, and the failure of the product – in spite of the innovation and creativity involved – was overwhelming. In retrospect, the reason for this failure seems clear enough. The innovation embedded in it benefited a population that would not purchase it (mainly non-smokers). The product appealed to the consideration of the user toward his environment, and an appeal to a "need of the second order" (the first order is, of course, the need to smoke) is always problematic.

To further clarify the distinctive nature of the Displacement Template here is another case. In the 1960s, the Ronson Company, producing gas lighters, introduced the idea of "a waxless candle" [see 2]. This candle was a shiny gas device easily ignited by a mechanism similar to that of a gas lighter, with a flame that lasted a long time. In addition, the flame of the new candle could be adjusted, which was a technological advantage over the traditional candle. The Ronson people believed this to be the key benefit, and it was at the center of the publicity campaign when introducing the candle.

In spite of its name (Waxless candle®), this innovation does not represent the Displacement Template, since the new product was not based on removing a component from the configuration of a candle, but on a new technological development of a lighter.

Consumers were not fascinated by the "new candle." The gentle flame produced by a wax candle affords a feeling of warmth and intimacy, not like a cold "techno-flame." Would we light a metal candle on a birthday cake? Or have a romantic dinner in its light? Marketers, blinded by the strong flame of the new candle perhaps forgot the emotional effect of the traditional candle. This may be one of the reasons for the failure of the product.

Displacement of a quantitative attribute.

Sometimes the reduction can be applied on a more quantitative level in which the component is not excluded, rather, a reduction is applied on one of its parameters. Consider the following example: Suppose that a careless truck driver is stuck under a bridge with his heavy truck loaded with giant containers. The containers are so large that they cannot be off-loaded, and the driver cannot move backwards or forwards. A possible solution may be reducing the air pressure in the tires and thus reducing the total height of the truck. This idea clearly has its limitations, however in cases of new products it may sometimes be appealing for marketing as discussed in the following examples.

1. Sales catalogs sold in the US are usually very bulky, in order to accommodate the maximum information. When a catalog of much smaller dimensions than usual appeared on the market, the scope of its sales surprised all marketing experts in the field. The developer of this catalog had noticed that bookshops stack these catalogs in piles. A small catalog, therefore, has a clear advantage: because of its small size it cannot be placed at the bottom, as the pile would then be unstable (see Figure 7.5.) The booksellers thus place it on top of the pile, and the customer's first encounter is with the small catalog. This change caused the sale of this catalog to leap upwards.

2. At the beginning of the twentieth century, a fault in the production line caused the appearance of an air bubble at the center of the bar of the Procter & Gamble's Ivory® soap. In an honest marketing move, the company

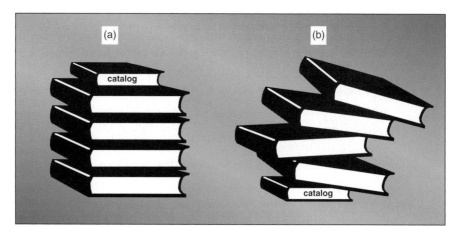

Figure 7.5   Marketing advantage of a small sales catalog. (a) The catalog at top of pile. (b) The catalog at bottom of pile makes it unstable.

announced that this was its fault, and that it would replace all the faulty bars.

When a pile of the faulty bars was accumulated in the factory, an employee noticed that this soap bar has a distinct advantage over others – it floats in the bath water due to its low density. Thus, when the bather drops it into the water, the time needed to retrieve it is much shorter, and it therefor dissolves less than a normal bar. It is, then, a thrifty and handy product. In fact, this Ivory® product was very successful throughout the century, with very little alteration.

In both examples the removal of a component was partial and concentrated on quantity (less material). Such a Displacement introduces the advantages of reduced costs and R&D while offering new opportunities. We note again that the difference between Displacement and unbundling is the fact that the reduced density soap was not sold at a lower price because of less ingredients. Rather, it was sold better because it floats. The floating was actually a new benefit of the reduced product.

Operational prescription

Operational prescription
1. Make a list of internal components (over which the manufacturer has control).
2. Make a list of external components (over which the manufacturer has no control).

3. Construct a product configuration mapping all desirable control connections between all listed components.
4. Locate the essential components, mark them and list the functions they fulfill in all their connections.
5. Choose an essential component and exclude it from the configuration, along with the function it had fulfilled.
6. Search for a market benefit resulting from the displacement you have effected. Remember that the principle Function Follows Form supports you. Try to find new market niches that may derive benefits from the removal.

REFERENCES

1. Mingo, J. (1994) *How the Cadillac Got Its Fins.* New York: HarperCollins Books.
2. Adler, B. and Houghton, J. (1997) *America's Stupidest Business Decisions.* New York: William Morrow and Company Inc.

8   The Component Control Template

Art does not reproduce the visible; rather, it makes visible

Paul Klee

What is the Component Control Template?

Before we examine the Component Control Template, let us recall the example of the Post-It Notes® (see more details on this case in Chapter 1). The development of that product shows that the idea of the "weak glue" on the notes emerged from the process of development of glue. This glue matches the Attribute Dependence Template – it is weak under bending stress but rather resistant to shear. However, the brilliance of the idea is not in the glue itself but in its connection to a colored memo slip, of the type used on an office desk.

Such small slips on which notes and memos may be written existed before the development of Post-It Notes® – in fact, that was the basic product from which the sticky notes were developed. However, the original product was characterized by a somewhat problematic attachment to the surface on which it was placed: it could blow away with every light wind and disappear. Note that the table is an external component to the Post-It Notes® (not controlled by the manufacturer), yet it is an important component in the configuration of the note.

Reminder: An external component is one that comes in contact with the product at a certain point in time, but is not controlled by the manufacturer.

The change in the configuration of the product that should enable the connection between the product and the external component was based on the extra link introduced between the surface and the note. In the original configuration, the surface placed the note at a certain height (see Figure 8.1a). In the

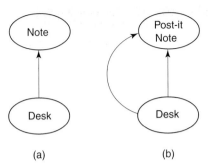

(a)   (b)

Figure 8.1   Operation of Component Control Template on a slip of paper, creating the configuration of Post-It Notes®. (a) Product configuration of a note. (b) Product configuration of a Post-It Note®.

new product, an extra link was formed between the note and surface by the introduction of a new internal component into the configuration – the glue. The desk now affixes the note at the horizontal level as well, preventing it from sliding along the surface. Figure 8.1b represents the new product configuration.

Another example of the use of the Component Control Template is a radiation-filtering computer screen. This screen was developed to solve a problem that arose through the continuous use of a computer: the radiation emanating from the computer caused the user headaches, eye strain and dizziness.

The negative effect of the electromagnetic radiation emanating from the product and the external component in the configuration of the computer (the users' eyes) represents another set of cases consistent with the Component Control Template. The external component is not controlled by the manufacturer or the deliverer of the service. The solution for this negative effect is to include a component with a new link into the configuration, which limits the amount of radiation. Figure 8.2 describes this negative effect, and the solution of including the new component in the product configuration.

In order to clarify the operation of the Component Control Template, let us observe the evolution of another product – hair shampoo. In the past, hair was washed using soap flakes. These flakes have a high rate of dissolution and create an extensive surface area that comes in contact with the hair. They adhere to fatty dirt, and when the hair is rinsed, they are rinsed off along with the attached dirt. The basic product configuration is shown in Figure 8.3.

The first shampoo was developed by Unilever, based on a liquid soap that came in contact with a greater area than did the solid soap. With the success of this product an independent product category developed, known to us as

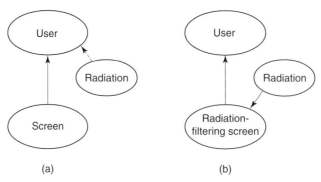

(a)   (b)

Figure 8.2   Analysis of radiation-filtering computer screen on the basis of the Component Control Template. (a) Computer radiation is an external component with a negative connection to user. (b) The radiation-filtering component solves the negative connection.

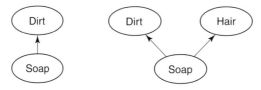

Figure 8.3   Product configuration of basic shampoo.

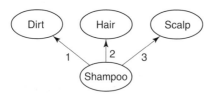

Figure 8.4   Product configuration of dandruff-preventing shampoo. The numbers of the links represent the order in which they appeared.

shampoo [1]. This category was developed in various versions. The basic shampoo product possessed specific characteristics, responding to different aspects of the consumer environment. Thus, for example, one type of shampoo was developed for the treatment of the scalp (preventing dry skin and dandruff). Figure 8.4 describes the product configuration of a dandruff-preventing shampoo (note the new link between shampoo and skin of scalp).

With the evolution of shampoo, this product responded to other, newer needs: destroying lice in hair, "tearless" shampoo, etc. These products have a new link of control with an external component that was not previously included in the product configuration (i.e., lice, or sensitive eyes). Figure 8.5 describes the product configuration of these shampoo products. The external components were previously present in the environment of the product, but

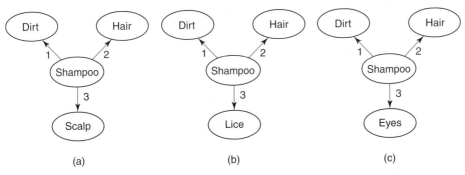

Figure 8.5   Configuration of other shampoo products. (a) Anti-dandruff. (b) Lice-killing. (c) Tearless.

were not included in its configuration. The new link between the external and internal components was achieved by the addition of a new component – killing lice or preventing eye irritation. Note that in the course of its development, the product "creates" a new link with its immediate environment. It "reaches its hands" to new environmental components in order to respond to problems they create. We therefore use the term "component control" because it involves a control, which frequently implies separation, blocking, or at least limiting the impact of one component over the other. As shown in the examples, a "link" does not necessarily imply uniting or bringing components closer.

> The Component Control Template is characterized by making a new link between a component in the internal environment of the product and a component in its external environment.

We must note that an "environment-adapted" shampoo such as the dandruff-preventing shampoo requires R&D and new technological setups. This is a frequent characteristic of the Template: adaptation of the product to the environment often requires investment in research and technology to a larger extent than the other Templates we have discussed. However, there are cases in which the technical alterations needed for adapting the product to its environment are minor, and all that is needed is the formulation of a new marketing message. We shall elaborate on this point later on.

The thought process inherent in applying the Component Control Template

The Template is based on the identification of a negative connection between an external component and the product configuration. This connection is solved by establishing a new link between the external and internal components. The thought process in applying this Template is initiated with the selection of a component external to the product, and then with an effort to locate a problem embedded in the existing connection between the selected component and the product configuration. The assumption behind this is that it is easier to locate problems than to search for solutions to needs.

We recall that the definition of the term "external component" confines us to a limited list of components. Their number is finite and they are easy to define, being in direct contact with the product. This procedure keeps our thoughts from wandering, and enables us to define the given space of possibilities for innovation and to scan it systematically.

Let us follow the process of implementing the Component Control Template on hair shampoo, in order to arrive at an "already-invented new" product. We shall first ask what external component comes in contact with hair shampoo. A partial list would be hair, scalp, water, hair conditioner, regular soap, towel, body, dandruff and sunlight. The sun's rays are the external component par excellence. Although the shampoo is rinsed off and apparently does not come in contact with sunlight, some components of it in fact stay on the hair, performing different functions (hair care, strengthening, conditioning), and come in direct contact with the sun's rays. Note that defining the sun as an external component is the result of careful scanning of all environmental components external to the product.

Having identified an external component that comes in contact with one of the internal components, we try to find a benefit that may result from connecting it to one of the components in the configuration. Scanning of benefits is assisted by analyzing the potential control between the external and internal components, and trying to locate a negative state (a problem). We can define the new benefit as a result of limiting or neutralizing the negative state. Note that locating negative states is comparable to searching for needs, yet it is easier and more accessible to our thought process. Is there a problem, for example, in the contact between sunlight and hair that is a component in the configuration of shampoo? If such a problem exists, the contact between shampoo and sunlight must be such that it will limit this problem or prevent it altogether.

You have probably already identified the problem in the contact between the sun's rays and hair: the damage to hair from solar radiation. The shampoo must present a new (physical) connection to sunlight limiting this damage. This definition means that radiation-filtering substances should be added to

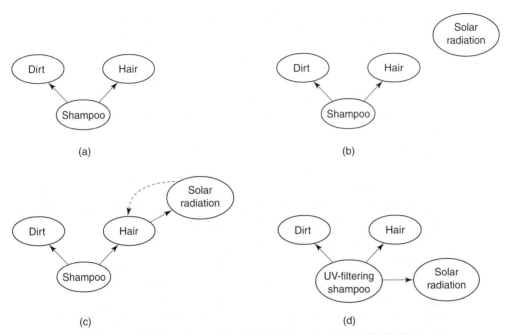

Figure 8.6   Operation of Component Control Template and construction of configuration of UV-filtering shampoo. (a) Product configuration of basic shampoo. (b) Locating an external component. (c) Identifying a negative connection between external component and product configuration. (d) Solving the problem that underlies the negative connection by addition of new component to product configuration.

the shampoo. These substances must remain in the hair after rinsing, and must therefore include the following characteristics:

1. Low mixing in water.
2. High absorbency in hair.
3. High degree of filtering.

Figure 8.6 describes the thought process of constructing a new product configuration of shampoo that filters ultraviolet (UV) radiation from the product configuration of basic shampoo:

1. Construct the product configuration (Figure 8.6a).
2. Include an external component in the configuration environment (Figure 8.6b) and search for problems resulting from its inclusion in regard to other components. Problems located between the external component and internal components are marked with a broken line, and termed "negative connections" (Figure 8.6c).
3. When we find such a connection, we must define a new functional link between an internal and an external component, responding to the problem represented in the figure (Figure 8.6d). If no internal component

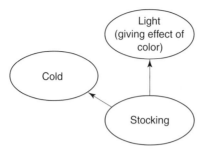

Figure 8.7   Configuration of women's stockings.

is able to make the necessary link, we then search for a new component to be included in the configuration.

Component Control without the need for a change in the product

Sometimes, Component Control does not require the addition of a new component to the configuration, but makes do with exposing an unknown aspect of an existing component or of the configuration itself. The control in these cases is not physical, but is expressed by rephrasing the marketing message of the product or its components.

Let us look at women's stockings. Here, too, UV radiation is considered an external component, as the legs are exposed to radiation dangers. Suppose that the awareness about radiation damage is high, and the creation of a benefit of filtering the radiation is relevant and welcomed by the market. If we produce a basic product configuration of a stocking (Figure 8.7) and apply to it the model of Component Control, we may arrive at a suggestion for stockings with a capacity to filter out UV radiation. Seemingly, all we have to do is add a radiation filter to the stockings, and announce stockings preventing radiation damage.

The announcement of stockings preventing radiation damage may help in creating an advantage over the competitors, and in increasing our market share by increasing the benefits of the product. See Figure 8.8. But we must first examine whether this is feasible from a technological point of view. As a matter of fact, even material as thin as nylon stockings filters more than 80 percent of UV radiation. It appears that the stockings already embody a hitherto unexploited formula of Component Control. The product is, in fact, already "adapted," and the only thing lacking is a marketing move, announcing the (true) character of the stockings and taking advantage of the function that the stockings provide free of charge.

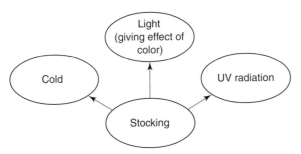

Figure 8.8   Component Control affords a new configuration to women's stockings, although no change has been made to the product.

This example demonstrates the model of Component Control. Some cases require R&D effort in order to create a new product (e.g., UV-filtering shampoo), while others may not require any technical change, so that R&D efforts are minimal. In the latter case the effort associated with the Component Control is confined mainly to marketing.

Another problem arises concerning women's stockings worn in summer. One of the external components of stockings is sweating, causing discomfort to the wearer. This is, in fact, the main reason that stockings are a seasonal product in warm countries. We need to examine whether one of the components of stockings may serve as a (positive) connection that will solve this problem. As the stockings cannot prevent sweating, we can only consider the possibility of conducting the sweat away from the skin. There are two basic possibilities for such action: (1) the stockings will absorb the sweat; (2) the stockings will drain the sweat outward, thus enhancing its evaporation.

The first possibility is not feasible for many reasons. Beyond considerations of comfort and esthetics, cooling occurs only when the sweat evaporates. The only feasible solution, then, is draining the sweat. The technical question of how to create such a link could be turned over to the appropriate professionals, as there seems to be a need for extensive R&D effort. Yet, the reverse is actually true. It seems that in some stockings, the mesh of fibers has a capillary effect (by which liquid in a long and narrow space is sucked up even against gravity). Such stockings draw the sweat toward their outer layer, where it evaporates and cools the legs. Cooling the legs by means of summer stockings is a substantial benefit and a powerful marketing message: it is now advantageous to wear stockings in summer. Note that although technically there is no need to make changes in the existing product, it is necessary to form a launching policy and extend marketing activity as required for any new product. Summer stockings are now available in the market (e.g., under the brand name Cacharel®).

Orbit® chewing gum: a medical discovery or a brilliant marketing move?

Orbit®, Tident® and other chewing gums with dental care promise hit the market like lightning, gaining a large market segment in a short period of time. The product was advertised as having a unique innovation: it aids in preventing dental caries. The Dental Association confirmed its effectiveness and, carried by the wave of health foods, it quickly became a standard health product.

Actually, the "technological" innovation in the Orbit® chewing gum is not so revolutionary. Every low-sugar chewing gum decreases acidity in the mouth in a way that delays the onset of caries. There is neither a new component nor a new function in Orbit®. It is a regular low-sugar chewing gum, whose success is the result of a brilliant marketing move. This is an additional instructive example of the Component Control Template, in which a product displays a marketing innovation requiring a marketing campaign only.

Observations on the Component Control Template

Component Control is a Template that operates in the market following certain dynamics, enabling the product configuration to send out tentacles to its environment. Before we discuss the characteristics of this dynamic and the way in which we may see in it a signal from the market about the possibility of using this Template, let us comment that this dynamic does not contradict our fundamental requirement that information is embedded in the product itself. The signals from the product environment help us determine which Template to use for developing the product but they do not assist us directly in developing a new idea or product.

The Component Control Template, connecting between the product configuration and components in the external environment, creates two marked effects:

1. An internal effect: The attention of the R&D team of the company is shifted from the main benefit to secondary benefits of the product, and resources are allotted to the development of these benefits.

2. An external effect: The market has the impression that the product offers the main benefit in an optimal way, and that the technological capability of improving upon the main benefit has been exhausted.

Operation of the Template may be problematic when the market does not yet feel assured that the product can actually supply the main benefit at its best, and prefers a competing product seen as focusing on this main benefit. When the market feels that the main benefit has not been exhausted, it is better to focus on the development of this benefit, using the Attribute Dependency Template for example. Attribute Dependency makes an internal analysis of the variables of the product using a predictive matrix, and enables its development in a way that serves the benefits that already exist in it.

The Component Control Template must be operated only when conditions enabling it appear in the market – mostly a feeling of trust by the market in the ability of the product to provide the main benefit. This trust, expressed in the stabilization of consumption habits of the product, is essential for presenting new secondary benefits in the product configuration, or in operating the Component Control Template. The secondary benefits may later develop into a separate product category, as was the case of the shampoo, or as often happens when the market is free to focus on a new benefit.

Operational prescription

Operational prescription
1. Make a list of internal components (over which the manufacturer has control).
2. Construct a product configuration: mark all the control links existing between the identified internal components.
3. Make a list of environmental components that come in physical contact with the product configuration.
4. Scan the environmental components one by one, trying to locate for each a negative connection – existing or potential – with the product configuration.

Note: since the number of environmental components in contact with the product is limited, systematic mapping may cover all possibilities of Component Control.

REFERENCES

1. Goldenberg, J., Mazursky, D. and Solomon, S. (1999) "Creative sparks," *Science*, 285 (5433), 1495–1496.

Part III

A closer look at Templates

It should not be a surprise if, when we examine other fields in which creativity plays a crucial role, we find evidence for the universality of Templates. If Templates indeed represent replicable patterns that may be generalized across variables and products, we should expect to discover evidence of Template dynamics in other ideation and problem-solving domains. Consequently, when this research paradigm of "idea archeology" was applied to advertising (perhaps as in other domains of arts) creative activity was found to be of prime importance. In Part III we present the implications of Templates on creative executions of advertisements. Here too, successful ads were typically found to share certain specified abstract patterns. The theoretical rationale for the emergence of such Templates is similar to that of the formation of Templates in new products. Chapter 9 describes and illustrates the Template theory in the context of advertising.

In addition, at this stage of developing the Template approach it is useful to outline it as a specified, formal framework. At various points in the book we have provided configurations of products to exemplify how well defined they are and how they can be used for creativity tasks in a relatively straightforward manner. In Chapter 10 the inference and development of the Templates is presented as part of a formal presentation that can also assist in generalizing and applying them in a variety of other contexts.

9   Templates in Advertising

Introduction

Among all the domains in which problem-solving paradigms are applied, marketing is most probably the most dominant in its attention to and respect for innovative ideas. This salience ascribed to innovativeness in marketing may be a consequence of the fact that in some situations (e.g., new products) innovation provides added value. For example, an engineering solution is selected when its cost-effective value meets predefined requirements, a scientific theory dominates because it is correct, while the effectiveness of an original idea in marketing sometimes may provide a competitive edge beyond cost–benefit considerations. Consequently, marketing presents an ideal arena in which ideation, problem-solving and innovation may be studied and in addition it affords plenty of documented evidence.

The advantages of the Template approach lie both in its consistency with related theories on ideation and in its practical utility. On the theoretical plane, Templates which are generated by analyzing the evolutionary trends of successful products at their mature stage are applied to non-mature situations. The empirical findings reported later in this book indicate that ideas elicited by individuals trained in Template ideation were judged superior to ideas produced through alternative methods, such as morphological analysis and random stimulation.

It should be expected, therefore, that by examining other fields in which creativity plays a crucial role we can find evidence for the universality of Templates. If Templates represent replicable patterns that may be generalized across variables and products, we should expect to discover evidence of Template dynamics in other ideation and problem-solving domains. Below we present the implications of Templates on creative executions of advertisements. The theoretical rationale for the emergence of such Templates was presented in Goldenberg, Mazursky and Solomon [1] and it is similar to that of the formation of Templates in new products emergence. As in the former context, the Templates were found to be identifiable, objectively verifiable and generalizable across multiple categories.

The fundamental Templates of quality advertisements

Creativity in advertising frequently involves methods that encourage the generation of a large number of ad concepts [2], on the assumption that the rewards of producing a large number of ideas will outweigh the costs [3]. The generation of new ideas in this manner tends to be highly unformalized and unsystematic. Often, such methods are based on the *divergent thinking* approach [e.g., focus groups, free association and other projective techniques; see 4] whereby judgment is suspended and ideas emerge by associative thinking in a "limitation free" environment.

However, even in a divergent thinking context certain patterns of creativity may emerge and be applied. Creative teams often seek ways to become more productive as they progress from one creativity task to another. *Common patterns* relevant to different domains are sometimes identified. These may then be applied on an ad hoc basis within a given advertising context, or even transported to other contexts. Such patterns will be more stable and less transient than the abundance of random ideas that emerge in the process of divergent (e.g., associative) thinking. They may also help in "organizing" the creative process by promoting routes that have been proven to lead to productive ideas and avoiding those that do not. Nonetheless, even if they prove productive, such patterns tend to be idiosyncratic and they are often not verbally definable. As such, they are likely to lack permanence and generality. We posit that certain patterns can be generalized to fit our notion of Creativity Templates, and that these Templates underlie the generation of quality ads. Because they can facilitate focused creativity, they will lead to more effective outcomes.

Let us portray these notions with an example. Figure 9.1a shows an ad for the 1989 French Open Tennis Championship sponsored by Penn [5]. The ad features a croissant-shaped tennis ball or, viewed differently, a croissant with a (Penn) tennis ball surface. The pattern of this ad served in generating another featuring a hockey puck shaped tennis ball for the Canadian Open Tennis Championship [Figure 9.1b, from 6] and a moisture ring of a tea cup for Penn's sponsorship of the British Lipton Tennis Championship [Figure

(a)

(b)

(c)

Figure 9.1   Examples of Replacement-based ads for tennis tournaments. (a) French Open Tennis Championship.
(b) Canadian Open Tennis Championship. (c) British Lipton Tennis Championship.

9.1c, from 7]. The common pattern in all three ads is a modified tennis ball
designed to symbolize a country. Consistent with the original ad and its principle pattern the US Open Championship might consider using a ham-
burger-shaped tennis ball or the American flag with tennis balls replacing the
stars.

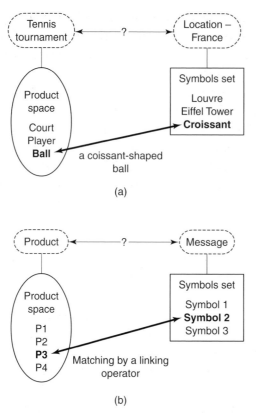

Figure 9.2   (a) Specific scheme underlying the French Open Tennis Championship ad. (b) General scheme underlying the replacement version of the Pictorial Analogy Template (discussed below).

Identification of the Creativity Templates

The common pattern of the tennis ball ads can be depicted schematically. From analysis of the French tennis tournament ad (Figure 9.1a) a scheme depicting the possible links between the tennis tournament and France can be constructed. Figure 9.2a provides the breakdown of the tennis tournament (left hand side) into some of its internal components such as a player, court or ball. The message theme in this ad (right hand side) is the location which in the case of France, can be represented by various symbols such as the Louvre, the Eiffel Tower or a croissant. The advertised event and the main theme (France as location) are then unified by matching their shapes. In a similar manner, schemes can be constructed for the ads which stress the importance of other locations. Note, however, that these schemes depict specific combinations of events and locations.

Generalization of this operation may be achieved by inferring a general scheme. A scheme can be considered general only to the extent that it can be widely applied in a variety of products, events and messages. The repetitive appearance of a scheme in different domains reveals the Creativity Template. Thus, the transformation from a specific scheme (Figure 9.2a) to a general scheme (Figure 9.2b) extends the notion of *common patterns* to the notion of Creativity Templates. You may recognize this scheme as the Replacement Template in advertising that was presented in Chapter 6. The general scheme, shown in Figure 9.2b, consists of two parts. The first part, denoted as the *product space,* is formed by the internal components of the product and the objects that interact with it (P1, P2, P3, P4 in Figure 9.2b). The tennis ball is a major internal component of the set of components that feature in a tennis championship. The second part, denoted as the *symbols set,* is formed by the symbols that feature in the consumer's representation of the message. In the tennis tournament examples, the croissant, puck and tea ring were chosen to symbolize the countries.

The elements chosen from the two parts of the general scheme (the product space and the symbols set) are then unified through a linking operator which matches their shape, color or sound (recall the illustration of the linking operator in the Nike-Air® ad; see Figure 6.20) Note that the product space and symbols set and the specification of their matching mechanism can be traced in additional domains. Although at first glance it may appear remote, the ad for Nike-Air® sneakers has the same fundamental scheme as the tennis tournament ads. Figure 9.2a depicted the scheme underlying the Nike-Air® ad. Replacement is obtained when a product (e.g., sneaker) or one of its parameters, replaces a symbol consistent with the meaning of the conveyed message (e.g., the fireman's net). Conceptually, the Nike-Air® application is more abstract than the tennis tournament applications which involve simple duplication of a common pattern. The general scheme no longer involves identical information, nor does it necessarily involve the same product. Yet, it is identifiable, objectively verifiable and generalizable across different ads. As such, it is defined as a Creativity Template.

In Goldenberg, Mazursky and Solomon [1] six major Creativity Templates were derived by inference from a sample of 200 highly rated ads. Judges found that 89% of the ads could be explained by the six Creativity Templates. A study comparing 200 award-winning and 200 non-award winning ads found that the two groups differ systematically in the number and distribution of Creativity Templates: 50% of the award-winning ads as opposed to only 2.5% of the non-winning ads could be explained by the Templates. In a study examining the robustness of the Templates, individuals were trained in Template-based idea generation, in an association technique, or not trained at all, prior to an ad ideation task. Another group later rated the ideas. Findings indicate that a priori knowledge of the Templates was associated with creating higher quality ads in terms of creativity, brand attitude judgments and recall, although the Template-based ads were found to vary in triggering emotional responses which included humor, sadness and annoyance.

Approaching creative advertising

The concept of structured creativity is already embedded in a number of current techniques such as morphological analysis. Other approaches, such as resonance and those involving the Janusian concepts, are more directly tailored to the context of advertising creativity. The resonance approach involves dual or multiple meanings surrounding a single word or phrase, such as "forget-me-knots" in an ad showing men's ties arranged to form a floral bouquet [see 8]. The Janusian approach involves "the capacity to conceive and utilize two or more contradictory concepts, ideas or images simultaneously" [see 9, p. 195]. Blasko and Mokwa [10] cite: "We're first because we last," and "We've got the inside of outside protection." Although (unlike the Creativity Templates described here) these methods do not lend themselves to schematization, they do provide important rules for creativity. An interesting step toward generating a broader taxonomy of figuration modes was recently presented by Mcquarrie and Glen [8]. They propose a rhetorical perspective contending that the manner in which a statement is expressed may be more influential than its content. Finally, it is interesting to note a specific area of advertising, namely humor. Certain dimensions of ads were found as successful predictors of humorous ads. Alden, Hoyer and Lee [11] used Raskin's psycholinguistic theory of humor to explain why certain ads are perceived as more humorous than others. They found that ads that employed a contrast between everyday life and the unexpected were generally perceived as more humorous than those employing a contrast between everyday life and the impossible. By extending these approaches, that typically emerged from a specific pattern of creative execution, the Creativity Template taxonomy provides fundamental generalizable structures for the generation of quality ads.

In reviewing the relevant research, one should assess the distinct contribution of several well-known taxonomies of advertising strategies focusing on other perspectives of ad generation, vis-à-vis the Creativity Template approach. For example, Simon [12] proposed a framework which includes ad strategies such as information, argument, repeated assertion, command and symbolic association. Similarly, Burke et al. [13] proposed a taxonomy in which positioning (defined as the featured benefits and the distinctiveness of a brand relative to other brands) and message emotion (the emotional tone of the ad) are key ad strategies. The main distinction between these frameworks and the Creativity Template taxonomy lies in the level of cognitive representation of the framework factors. Whereas the advertising strategies represent summative factors and intended consequences (e.g., emotional response), the Creativity Templates represent the scheme that antecedes and gives rise to these strategies. A specific, well-defined Template may evoke an emotional response, but the emotion itself neither offers the scheme nor the means for eliciting this response. The aforementioned advertising strategy frameworks thus focus on the decision between different consequences (e.g., emotion, positioning), while the Creativity Template approach focuses on the cognitive activities leading to these consequences.

Demonstrating Templates in advertising

The universal nature of the Template paradigm can be illustrated by exploring the field of advertising, as discussed below. More details on this research can be found in Goldenberg, Mazursky and Solomon [1].

Inferring Templates

The search for Templates was conducted by inferring the linking operators in each ad and identifying the relevant sets and their spaces, in a manner consistent with the inference of the Template described above (known as the Pictorial Analogy Template). For example, an ad showing a lady barking at burglars and scaring them off in an advertisement for a security lock (see Table 9.1), led to the identification of a Template version, termed Absurd Alternatives, by the following procedure. The product (lock) and the message theme (security) were identified. Then the linking operator was inferred – the lady's voice was replaced by dog barks. An unanswered question was "what elements are linked by the linking operator?" At one end of this link, the lady being threatened by a burglar is a situation which provokes the need for security (the message). At the other end, the dog could serve as an alternative to a lock in enhancing security. Other options serving the same function

Table 9.1 Examples, descriptions and formulation of Template versions

	I. The Pictorial Analogy Template: the Replacement version	II. The Extreme Situation Template: the Absurd Alternative version
Example	Examples and detailed formulation of the scheme underlying the Replacement version of the Pictorial Analogy Template (composed of a symbols set, a product space and a linking operator) are presented in the introduction to this chapter, in elaborating the notion of Creativity Templates (see Figures 9.1–9.3).	The commercial for locks showing an old lady scaring away burglars by barking at them (Suissa Miller Advertising Company, USA, 1993, Cannes contest award) conveys the message that a safe and peaceful evening can be achieved either by buying a certain lock or by barking.
Description	See description in the introduction to this chapter.	The idea of this version is to present a tongue in cheek suggestion to the viewer: "You don't have to b our product. There are *alternative options* for achievin the same results, such as . . ." The *alternative option* is presented in a seemingly serious manner but, contrary the declared position of the advertiser, the viewer will draw the conclusion that such an alternative is absurd ridiculous. The following elements typically appear in this version 1. An unexpected shift in the consumer's frame of mir into an imaginary status or into a different product category (or, unusually, to a competitive brand). 2. The absurdity and extreme unrealism of the alterna option are obvious and recognizable by the consum Any attempt to make the alternative more realistic would only weaken the claim of the ad.
Formulation	See formulation in the introduction to this chapter.	The specific scheme of the lock commercial consists of two sets. A *set of alternative options* and a se *situations*. An *alternative option* is an object or an actio dog, in this case) which can be used to achieve the product's attribute (safety). The alternative option doe not have to be realistic, although it is assumed that the target audience will be familiar with it. A *situation* is a common use scenario of the product in time and place our case a peaceful evening in the home of an old lady The linking operator links one element from the situat space (the lady) with one element in the alternative sp (barking).

le 9.1 (*cont.*)

	I. The Pictorial Analogy Template: the Replacement version	II. The Extreme Situation Template: the Absurd Alternative version
eral eme	See scheme in the introduction to this chapter.	
cific eme	See scheme in the introduction to this chapter.	

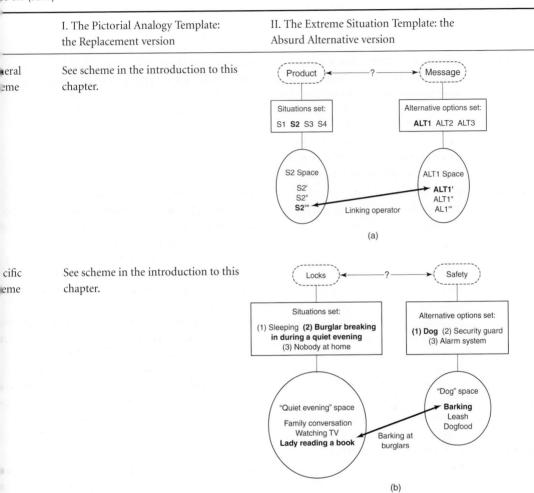

(a)

(b)

Table 9.1 (*cont.*)

	III. The Consequences Template: the Extreme Consequences version	IV. The Competition Template: the Uncommon Use version
Example	A commercial for car loudspeakers showing a bridge on the verge of collapse when the loudspeakers of the car parked on it are turned on at high volume. The message is that the music can be played so loud that even the sturdy foundations of the bridge are threatened by its impact (BBD, Los Angeles, 1994, Cleo award winner).	A commercial for jeans showing a couple in a broken-down car being towed by a pair of jeans tied to the rescuing car.
Description	The idea of this version is to present an extreme consequence of an emphasized product attribute. The absurdity of the consequence, even though presented in a serious manner, is eminently obvious to the viewer. Therefore, even a negative result (the collapse of a bridge) is conceptualized as an indication of the quality of the product. The following elements usually appear in this version: 1. Consequences based on a true fact: The extreme situation is created by taking a key attribute of the product to the extreme (e.g., the sound emitted by the loudspeakers causes objects – even a sturdy bridge – to vibrate). 2. The absurdity and extreme unrealism of the consequences are obvious and recognizable by the viewer.	The idea of this template is to emphasize a product attribute by applying it to solve a problem in a context totally different to its intended use. The following elements typically appear in this version: 1. A problematic scenario or issue. 2. Ambiguity as to the product to be the subject of the when the problem or dilemma is presented.
Formulation	The specific scheme of the loudspeaker commercial consists of two sets: a *set of situations* and a *set of consequences*. A consequence is a phenomenon, action or behavior which results from the product attribute appearing in the message. The consequence has to appear familiar and not unreasonable to the target audience (e.g., vibrations). It does not have to be absurd or extreme. The linking operator acts on the product and a selected item in the consequences set by taking the consequence to an extreme.	The specific scheme of the jeans commercial consists o two sets: a *set of situations* and a *set of problems*. The problem suspends the natural flow of events in the situation. The situation in our example is a couple driv a car. The problem is the break-down of the car. The viewer expects to see "how it is going to continue from here." The problem will be solved by using the product is therefore important to invent the problem by thinki "backwards" so that the product attribute contained in message will provide its solution. The link is the use of product as a solution by exploiting the attribute (the strength of the jeans).

le 9.1 (cont.)

The Consequences Template: the
reme Consequences version

IV. The Competition Template: the Uncommon
Use version

eral scheme

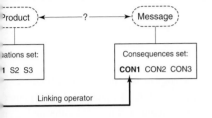

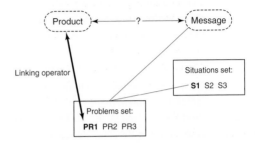

cific scheme

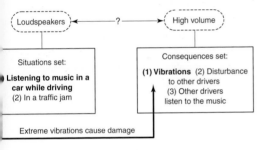

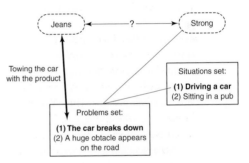

Table 9.1 (*cont.*)

	V. The Interactive Experiment Template: the Activation version	VI. The Dimensionality Alteration Template: the Time Leap version
Example	An example of the activation version is an ad containing a large black patch. When the viewer performs the action suggested in the ad, he/she will become aware of the necessity of an anti-dandruff shampoo (DDB, Needham San-Paulo, 1995).	A commercial for life insurance showing a wife arguing with her husband for canceling his life insurance. The whole scene takes place after he dies, and portrays the w communicating with her late husband in the setting of séance (a Cannes award winner in 1993).
Description	The consumer is required to perform a task or experiment in order to receive the message conveyed by the ad. The message is contained in the compelling result. Most of the ads in this category convey a message emphasizing a need or a problem which can be resolved if the product is used. The following elements typically appear in the activation version: 1 An experiment requiring *physical action.* 2 The experiment is executable on the spot. 3 The experiment's results highlight a general need rather than a unique quality of the specific brand.	The idea of this template is to present an ordinary situation (in this example, an argument about whether continue investing in the product). The entertaining eff is achieved by shifting the scenario to the past or the future.
Formulation	The specific scheme of the anti-dandruff shampoo ad consists of two different sets: the *senses* set and the *experiment* set. The relevant senses set is drawn from the list of the five senses. The experiment set consists of test scenarios to ascertain need for the product. The linking operator requirement is that the experiment represented in the experiment space will be performed *physically by interacting with the media* (newspaper, radio etc.).	The specific scheme of the life insurance ad consists of two sets: a *set of episodes* introducing the message claim (e.g., EP2 – a wife arguing with her husband) and a *tim set* (past, future). First, the episode space is selected (e.g wife, husband). Then an operator links an element from the time set and an element drawn from the episode sp (e.g., the husband's life status is transferred into the future). Note that the invented situation in the different time frame has to be relevant to the product and its attributes and, therefore, in this case, the future is more appropriate.

ble 9.1 (*cont.*)

The Interactive Experiment Template: e Activation version	VI. The Dimensionality Alteration Template: the Time Leap version

neral scheme

ecific scheme

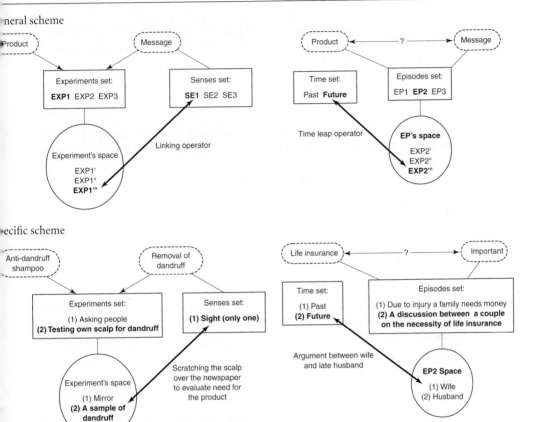

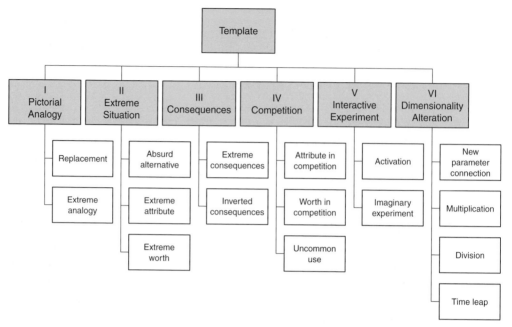

Figure 9.3   The Creativity Template taxonomy.

(e.g., security guards, alarm systems) are also available. Accordingly, the lady represents an element in a situation set, whereas the dog and the other options represent elements in an alternative option set. The Absurd Alternatives version of the Extreme Situation Template is obtained by the unrealistic nature of the solution created by the linking operator.

The repetitive appearance of this scheme in various ads dealing with different products and messages defines a general scheme, or a Template. A total of six key Templates and their sixteen versions were identified. A detailed description and formulation of the six Templates is given in Table 9.1, using one example for each Template. The following Templates and their versions were identified (see Figure 9.3 for overall structure).

I. The Pictorial Analogy Template.

The Pictorial Analogy Template portrays situations in which a symbol is introduced into the product space, as discussed in detail in the introduction to this chapter. This Template has two versions: the Replacement version and the Extreme Analogy version. In the Extreme analogy version the symbol is taken to the extreme whereas in the Replacement version it is merely transplanted.

II. The Extreme Situation Template.

The Extreme Situation Template represents unrealistic situations, in order to enhance the prominence of key attributes of a product or service. This category includes three versions: the Absurd Alternative version, the Extreme Attribute version and the Extreme Worth version. The Extreme Situation Template is exemplified and described in Table 9.1, using the Absurd Alternative version. The Extreme Attribute and Extreme Worth versions portray situations in which either the attribute or the worth of a product (or service) is exaggerated to unrealistic proportions (e.g., a jeep driving underneath the snow to demonstrate its all-weather driving ability – Cliff Freeman and Partners, NY, 1994).

III. The Consequences Template.

The Consequences Template indicates the implications of either executing or failing to execute the action advocated in the ad. The two versions of this Template are: the Extreme Consequences version (exemplified and described in detail in Table 9.1) and the Inverted Consequences version, warning against the implications of not executing the recommendation of the ad (e.g., an ad promoting a brand of vitamin, showing an otherwise highly energetic person unable to get out of bed in the morning). Figure 9.4 presents four examples of a possible execution of this template; one can see how the same theme can be replicated without diminishing the creativity of the outcomes.

IV. The Competition Template.

The Competition Template portrays situations in which the product is subjected to competition with another product or an event from a different class, where selection of the other product or event is guided by its expected superiority over the advertised product. For example: (1) a race between an advertised car and a bullet (Della Femina Travisanod and Partners, Los Angeles, 1990s) or (2) a person contemplating whether to continue eating the (advertised) cereal or to answer a ringing phone. There are three versions of the Competition Template: the Attribute in Competition version, the Worth in Competition version and the Uncommon Use version. The difference between the first two versions relates to whether the competition pertains to a product attribute or challenges the worth of the product. This Template is exemplified in Table 9.1 by the Uncommon Use version.

V. The Interactive Experiment Template.

The Interactive Experiment Template induces realization of the benefits of the product by requiring the viewer to engage in an interactive experience with

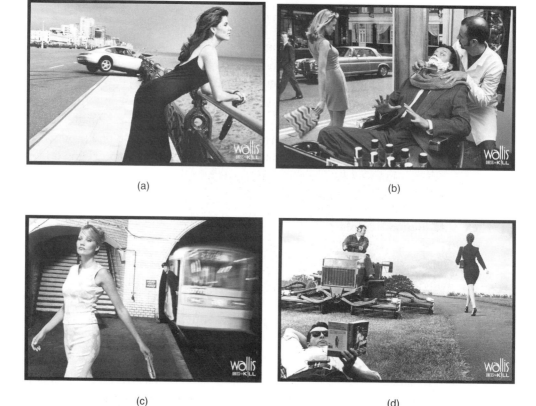

(a)   (b)

(c)   (d)

Figure 9.4   Examples of a shared Template in advertising (in this case: Extreme Consequences Template). (Bartle Bogle Hegarty, 1997.)

the medium in which the ad appears. This can be achieved by either actually engaging in an experiment (the Activation version – see Figure 9.5) or only imagining the performance of such an experiment (the Imaginary Experiment version). This Template is exemplified in Table 9.1 by the Activation version.

The notion of interactive experiment is different from the notion of "demonstrations" (often used by copywriters) in that demonstrations, despite their function in enhancing involvement, do not involve physical action in the manner described here.

VI. The Dimensionality Alteration Template.

The Dimensionality Alteration Template manipulates the dimension of the product in relation to its environment. Its four versions are: the New Parameter Connection version, the Multiplication version, the Division

Figure 9.5   Example of the Activation version of the Interactive Experiment Template. The text in the ad induces the viewer to "scratch his or her scalp over the black square and see whether s/he should read the text" (about an anti-dandruff shampoo).

version and the Time Leap version. In the New Parameter Connection version of the Template, previously unrelated parameters become dependent (e.g., the speed of a new aircraft is demonstrated by reducing the size of the ocean). The Multiplication and Division versions are executed by multiplying the product and comparing the duplicates, or dividing the product into its components and creating some form of relationship between them. This Template is exemplified in Table 9.1 by the Time Leap version.

Template distribution

Following Template inference, the distribution of Templates among the high-quality ads was examined. The Template ad classification was performed independently by two trained judges who each had at least 10 years' experience in the advertising field. They were taught to identify linking operators, and then given a list of possible spaces. Each Template was illustrated by five examples. They were then given an exercise testing their ability to correctly classify the Template. Each judge was asked to classify a set of ads (two ads per Template version, as well as non-Templates). A trained judge correctly classified more

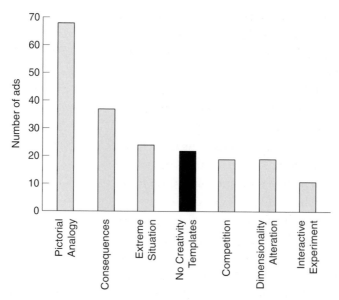

Figure 9.6   The distribution of ads by Templates.

than 95% of the ads in this exercise. The judges then classified the ads for this experiment. Inter-judge agreement rate in the assignment of ads into the six Templates was 94%, and disagreements were resolved by discussion. Of the 200 highly rated ads, 89% could be accounted for by the Templates. Figure 9.6 summarizes the distribution of the ads by their Creativity Template.

Implication of Creativity Templates on creative execution in advertising

Viewed from a micro perspective, individual advertisers may adopt idiosyncratic Templates and use them in generating new ideas. The research lends support to a broader perspective contending that the Templates may be widely applied across products, messages and target audiences. It serves to enhance the understanding of the emergence of quality ads as well as creativity in advertising.

The findings in Goldenberg, Mazursky and Solomon [1, 14] indicate the superiority of Template-generated ad ideas in creativity judgments, brand attitudes and recall. For example, the Interactive Experiment (Activation) Template requires physical activity and effort from the audience. Enhanced involvement is associated with high cognitive activity [15] and, thus, the Interactive Experiment is likely to be useful when the strategy is designed to cause a particular behavioral change which is less responsive to peripheral cues. The Absurd Alternatives and Replacement versions of the Pictorial Analogy and Extreme Situation Templates are more useful when humor is the intended strategy.

These findings may also be a step toward defining a comprehensive model of the antecedents of outcome reactions to advertising stimuli. Improved understanding of the wide spectrum of reactions connecting the basic Templates with end-user reactions is likely to be beneficial for both academics and practitioners. Such a framework would create a synthesis between the activity of creative professionals whose focal interest is generating the ads, managers whose main responsibility is strategy formulation, and the academic activity that focuses mainly on the consumer-reaction end of the advertising process. Obviously, further research is required to shed more light on this important aspect of Creativity Templates.

We may further postulate that the Template taxonomy provides the means to achieve "creativity expertise." Unlike the divergent thinking approaches in which the required expertise is not necessarily related to the creativity process itself (e.g., individuals can be trained to be better moderators in brainstorming), the Creativity Template approach lends itself to training and has the capacity to improve creativity outcomes directly. In fact, training individuals in Creativity Templates may result in higher levels of creativity expertise [16]. The Template taxonomy facilitates the focused cognitive effort involved in generating new ideas, the capacity to access relevant information, and memorability of the reduced set of information needed to perform the tasks.

In addition, the present investigation concurs with an emerging stream of research which de-emphasizes the traditional treatment of visual and verbal modes in advertising as functionally distinct entities. Some of the qualities of pictures which were believed, in the past, to characterize verbal information, and some of the qualities of verbal information which were previously more closely associated with pictures, are being re-examined. One direction of this research is visual rhetorics. Scott [17] challenged the assumption that pictures are merely reflections of reality, claiming that images represent complex figurative arguments. Although bearing on verbal information, Unnava, Agrawal and Haugtvedt [18] argued against the concentration of consumer research on visual imagery as the only type of imagery, claiming that words differ in the degree to which they provoke imagery or influence reading and listening. The Creativity Template approach is in accord with this research trend in that it treats the message and its delivery as a whole, rather than decomposing it into the functions carried by the visual and the verbal modes.

The Nike-Air® and the barking lady examples serve to illustrate the complex figurative arguments conveyed by pictures. Moreover, in the anti-dandruff shampoo ad, imagery provocation and informativeness are entirely independent of their verbal and visual components; neither the visual nor the verbal component can be understood as separate information entities, yet in combination they achieve a high level of imagery.

In creative thinking we seldom utilize even those regularities we have at hand. Considering the twentieth century alone, relational structures have been developed in a variety of disciplines, such as linguistics [19, 20], anthropology [21], random graphics [22], venture and transitional management [23], psychology [24] and artificial intelligence [25]. At least some of these are potential resources for inventive thinking, beyond serving as frameworks for historical organization.

One justification for examining regularities as a potential source of creativity is that there is already enough evidence for the universality nature of the Creativity Templates. Creativity perception may be enhanced because these structures match certain *attractors,* namely, paths that self-organized mind tends to follow.

These questions were addressed in a recent study. (Goldenberg, Mazursky and Solomon, [14] investigated a series of Template-based computer-generated ideas (in advertising) which were compared (by judges, in a blind experiment) to ideas proposed by subjects in four distinct control groups. Template-based ideas were judged to be superior to ideas proposed by individuals when evaluated on a number of scales of creativity, efficiency and effectiveness. The reported conclusion of this study identified superior "creativity value" inherently contained in the Template itself.

The computer was used to remove beyond any doubts the "human factor" in the Template-based ideation sessions. These results imply that the Templates' structures are not an illusion. They reflect some unique subconscious perception of creativity in our minds. It seems at this stage that, at least part of, these well-defined schemes are universal.

REFERENCES

1. Goldenberg, J., Mazursky, D. and Solomon, S. (1999) "Creativity Templates: towards identifying the fundamental schemes of quality advertisements," *Marketing Science,* **18,** 333–351.
2. Batra, R., Aaker, D. A. and Myers, J. G. (1996) *Advertising Management.* Englewood Cliffs NJ: Prentice Hall.

3. Winston, P. H. and Shellard, S. (eds.) (1990) *Artificial Intelligence at MIT: Expanding Frontiers.* Cambridge MA: MIT Press, pp. 430–563.

4. O'Guinn T. C., Allen, C. T. and Semenik, R. J. (1998) *Advertising.* Cincinnati, OH: South Western College Publishing.

5. *The One Show* (1989) Vol. 11. Switzerland: Rotovision, S.A.

6. *The One Show*, (1991)Vol. 13. Switzerland:Rotovision, S.A.

7. *The One Show* (1992) Vol. 14. Switzerland: Rotovision, S.A.

8. Mcquarrie, E. F. and Glen, D. (1992) "On resonance: a critical pluralistic inquiry into advertising rhetoric," *Journal of Consumer Research*, **19** (September), 180–197.

9. Rothenberg, A. (1971) "The process of Janusian thinking in creativity," *Archives of General Psychology*, **24** (March), 195–205.

10. Blasko, V. J. and Mokwa, M. P. (1986) "Creativity in advertising: a Janusian perspective," *Journal of Advertising*, **15** (4), 43–50.

11. Alden, D. L., Hoyer, W. D. and Lee, C. (1993) "Identifying global and culture-specific dimensions of humor in advertising: a multinational analysis," *Journal of Marketing*, **57** (2), 64–75.

12. Simon, J. L. (1971) *The Management of Advertising.* Englewood Cliffs, NJ: Prentice Hall.

13. Burke, R. R., Rangaswamy, A., Wind, J. and Eliashberg, J. (1990) "A knowledge-based system for advertising design," *Marketing Science*, **9** (3), 212–229.

14. Goldenberg, J., Mazursky, D. and Solomon, S. (1999) "Creative Sparks" *Science*, **285** (5433), 1495–1496.

15. Assael, H. (1998) *Consumer Behavior and Marketing Action.* Cincinnati, OH: South Western College Publishing.

16. Alba, W. J. and Hutchinson, J. W. (1987) "Dimensions of consumer expertise," *Journal of Consumer Research*, **13**, 411–454.

17. Scott, L. M. (1994) "Images in advertising: the need for a theory of visual rhetoric," *Journal of Consumer Research*, **21** (September), 252–273.

18. Unnava, R. H., Agrawal, S. and Haugtvedt, C. P. (1996) "Interactive effects presentation modality and message-generated imagery on recall of advertising information," *Journal of Consumer Research*, **23** (June), 81–87.

19. Eco, U. (1986) *Semiotics and the Philosophy of Language.* Bloomington, IN: Indiana University Press.

20. Chomsky, N. (1957) *Syntactic Structures.* S-Gravenhage: Mouton and Co.

21. Levi-Strauss, C. (1974) *Structural Anthropology.* New York: Basic Books.

22. Palmer, E. M. (1985) *Graphical Evolution: An Introduction to Random Graphics.* New York: Wiley and Sons.

23. Kauffman, S. (1995) *At Home in the Universe.* Oxford: Oxford University Press.

24. Simon, H. A. (1966) "Scientific discovery and the psychology of problem solving," in *Mind and Cosmos: Essays in Contemporary Science and Philosophy*, Vol. 3. R. G. Colodny (ed.) Pittsburgh: University of Pittsburgh Press.

25. Minsky, M. (1988) *The Society of Man.* New York: Simon and Schuster.

10 Further Background to the Template Theory

At various points in this book we have provided configurations of products to exemplify how well defined they are and how they can be used for creativity tasks in a relatively straightforward manner. In this chapter the inference and development of the Templates is presented as part of a formal presentation that can also assist in generalizing and applying them in a variety of other contexts.

Before describing the Creativity Templates themselves, let us revisit and elaborate the procedure for constructing the configuration of products. Several background definitions and rules of operation are required at the outset.

Space

Space is the field of operation. Templates operate in two spaces:
1. **The component space** consists of static objects – the fundamental component parts that make up the product as a whole, or the fixed external elements that have a direct impact on the product.
2. **The attribute space** consists of variables of the product or its components that can be changed.

Recall the chair example, the legs and seat of the chair would be components; the color and height of the chair would be attributes.

Note that the Creativity Templates approach considers only those attributes that consist of factual information. Abstractions or inferences, such as esthetics [see definitions in 1], are considered only at the later, development stage.

Characteristics

The characteristics of a product are its components and attributes. Characteristics may be either internal or external:

1. **Internal characteristics** are components or attributes that *are* under the manufacturer's control.
2. **External characteristics** are components or attributes that *are not* under the manufacturer's control.

In our chair example, two internal characteristics in the component space would be the legs and seat; two internal characteristics in the attribute space would be the color and height. Note that all characteristics of the product itself (product characteristics) are internal characteristics.

On the other hand, two of a chair's external characteristics in the component space would be the floor and a table; two external characteristics in the attribute space would be room temperature and the weight of the person sitting on the chair.

Rule: An object may be considered to be a characteristic of a product only when it is in direct contact with the product in its common usage. Thus, the floor and the weight of the person sitting on the chair would be considered legitimate considerations because they are components in direct contact with the product.

Links

A link describes the relationship between two characteristics and complies with the following rules:

1. A link must have direct influence.
    (a) A link may exist between two components only when a change in one component will be directly responsible for changing the parameters of the other component.
    (b) A link may exist between two attributes only when a direct dependency exists between the attributes.
2. A link must represent an assignment of function.
    The denoted influence between two characteristics must have been designated by the manufacturer.
3. A link must have consistency.
    Links may exist only between characteristics in the same space – components may be linked only to other components and attributes may be linked only to other attributes.

To remind ourselves of the concept of linkage, let us return to our chair example. Examine the links between the seat and the legs of a chair. There are two links, both in the component space of the chair. One link stems from the legs' support of the seat. Any change in the chair legs has a direct influence on

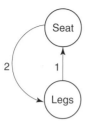

Figure 10.1   Link 1 – legs support the seat and hold it at the desired height. Link 2 – the seat holds the legs in place.

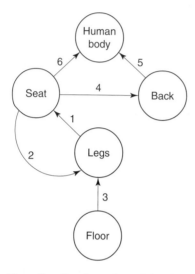

Figure 10.2   The configuration of an ordinary chair (oriented graph).

the seat. The seat is linked to the legs, too, in that it holds them in place (Figure 10.1).

Configuration

A product's configuration is defined as the complete set of the product's links and characteristics. As each product is different, each configuration is unique.

Configurations are simple to understand when charted by *oriented graphs*. Figure 10.2 depicts a configuration chart for an ordinary chair. Characteristics are denoted by circles and links by arrows. The direction of the arrow represents the direction of the assigned function.

Once a configuration is defined, its boundaries become fixed. Since all links and characteristics have been configured, there can be no other, alternative structure of the product without altering it in some way.

Note also that the configuration depends on both the product's structure (internal characteristics) and context of use (external characteristics).

Operators

The configuration of a product is like a "snapshot" of the product-based information. In order to create a new product, however, we must define the dynamics of change between past and future versions of products. Creativity Templates operate by defining the systematic changes between an early product configuration and the one that follows it. Those systematic changes are termed operators.

There are six elementary operators underlying the construction of the Templates. These are the means by which Templates operate across configuration boundaries.

Operators are one of the key elements that separate Creativity Templates from other taxonomies. Whereas a few other taxonomies have identified models of creativity patterns, they have not provided tools by which the patterns act to predict creativity. This is the function of the operators.

The inclusion and exclusion operators

The inclusion and exclusion operators import and export characteristics across configuration boundaries.

1. **The inclusion operator** imports an external component into the configuration.
2. **The exclusion operator** exports an *unlinked* component, either internal or external, out of the configuration boundaries (Figure 10.3).

The linking and unlinking operators

These two operators work on the links between characteristics.

1. **The linking operator** connects two unlinked characteristics.
2. **The unlinking operator** eliminates a link (Figure 10.4).

The splitting and joining operators

1. **The splitting operator** removes one characteristic (internal components) from the link. However, the function of the link is maintained (a dangling function).
2. **The joining operator** adds a (new) characteristic to a split link (Figure 10.5).

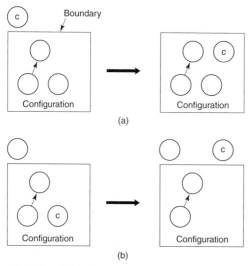

Figure 10.3   Illustration of (a) the inclusion and (b) the exclusion operators.

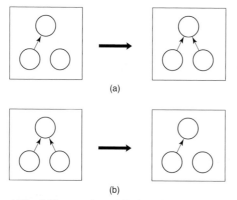

Figure 10.4   (a) The linking operator and (b) the unlinking operator.

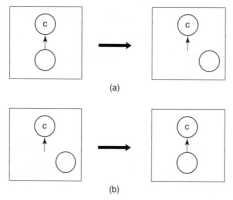

Figure 10.5   (a) The splitting operator and (b) the joining operator.

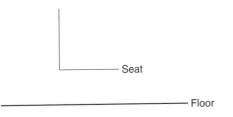

Seat

Floor

Figure 10.6   The legs of the chair are eliminated but their function remains.

The splitting operator creates an intermediate structure that may appear illogical and unrealistic. This inconsistency is a temporary structure in the abstract scheme. The link's function remains, however, and will be performed by a substitute component.

For instance, in the case of the chair configuration, a splitting operator could be applied to the legs–seat link, removing the legs. There would be nothing to hold the seat up once the legs were gone, but in our configuration, the seat would continue to be supported at the same height. Of course, this schematic is illogical and imaginary (Figure 10.6).

At this point, however, the removed component (the legs) would be replaced by an external component such as a wall or a table, which could also fulfill the desired function. The new component would be attached via a joining operator. In this case, we have actually developed the configuration for immovable chairs attached to the wall or the table.

Creativity Templates as macro operators

Transition from an existing product to a new idea can be accomplished by applying the above-listed fundamental operators in sequence. These "macro operators" constitute the Creativity Templates. We have identified five basic Creativity Templates: Attribute Dependency, Component Control, Replacement, Displacement and Division (the first four of these Templates were discussed in detail in Part II). Below we construct each Template as a "macro operator."

I. The Attribute Dependency Template

Attribute dependency is finding two **variables** that are not interdependent (i.e., a change in one does not cause a change in the other), and making a new connection between them. Recall the operational presentation of the Attribute Dependency Template (see Figure 10.7)

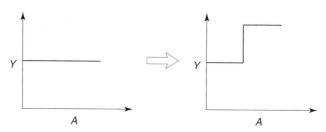

Figure 10.7   Attribute Dependency Template.

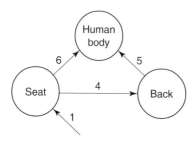

Figure 10.8   The chair configuration following splitting and exclusion of the legs.

The Attribute Dependency is the only Template that operates in the attribute space (all the others operate in the component space). The sequence of operators that constitutes this Template is *inclusion* and then *linking*, or *unlinking* and then *linking* two unlinked attributes.

II. The Replacement Template

This Template operates in the component space. An *essential* (internal) component is removed from the configuration. However, the link between the (removed) component and the other components remains. This generates a temporarily inconsistent abstract structure. Because of the dangling link, the operation is completed only when the missing component is replaced by another component. The replacement must be an external component with a function similar to that of the removed component. This Template (splitting, excluding, including, and joining operators) is described below and depicted in Figures 10.8 and 10.9.

1. **Split and exclude.** An intrinsic component is eliminated from the configuration while preserving its associated intrinsic function. In the resulting intermediate configuration, this intrinsic function is not carried by any component and remains as an unsaturated intrinsic function. The intermediate configuration is a necessary step in the Replacement procedure even though

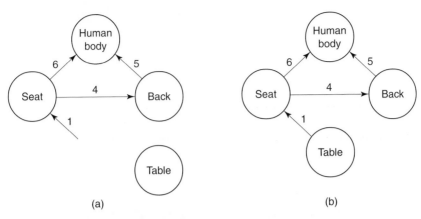

(a)   (b)

Figure 10.9   (a) Including table. (b) Joining the table to the seat.

it represents an incomplete product form. In this example, the chair legs are eliminated and their "holding" function remains to be performed by a new component. The resulting intermediate configuration is shown in Figure 10.8.

   2. **Include a suitable external component.** The unsaturated function can be fulfilled by an external component. As this component is out of the manufacturer's control, it has to bear a similar function. Such a component might be a table (Figure 10.9a).

   3. **Join.** By joining the incorporated component, a configuration of a new product is defined (Figure 10.9b). This new configuration creates an attachable chair, for example – a baby chair whose main benefits would be portability and keeping the baby at the appropriate height for any table.

III. The Displacement Template

Like the Replacement Template, the Displacement Template operates in the component space. Here too, an essential internal component is removed from the configuration. However, contrary to the former Template, its associated link is removed as well. In this case a new idea for the product must be based on a new appeal, one that the former product did not provide.

   The essential process of this Template is illustrated by the chair configuration depicted in Figure 10.10. The result of excluding the legs *and* their function may be a chair for use at the beach.

IV. The Component Control Template

This Template introduces a new connection between two previously unconnected components. The link in the Component Control Template can be

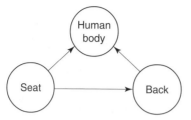

Figure 10.10   The Displacement Template. The chair legs are removed, along with their link to the seat.

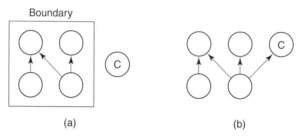

(a)   (b)

Figure 10.11   The Component Control Template. (a) Configuration before applying the Template and (b) after applying it.

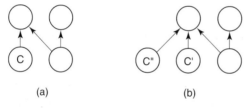

(a)   (b)

Figure 10.12   The Division Template. (a) Initial configuration. (b) Component C is split to give C′ and C″, each performing a different function.

with an *existing* (internal or external) component. The essential dynamics of the configuration are shown in Figure 10.11.

V. The Division Template

According to this Template, a component is split into two, and each new component becomes responsible for a different function. The outcome of this Template is shown in Figure 10.12.

REFERENCES

1. Chattopadhyay, A. and Alba, W. J. (1988) "The situational importance of recall and inference in consumer decision making," *Journal of Consumer Research*, 15 (June), 1–12.

Part IV

Validation of the Templates theory

This part of the book will present the validation of the Template theory and elaborate on some of its aspects. The presentation of the Templates alone may appear somewhat mystical to readers who may wonder about whether Templates prove superior in a more rigorous empirical examination, beyond the examples presented earlier. Understanding the theoretical grounds and the tests that challenge it should improve ability with and confidence in the implementation of the Templates. The empirical studies reported herein were published in Goldenberg, Mazursky and Solomon [1] and Goldenberg, Lehmann and Mazursky [2]. For readers unfamiliar with the statistical methods we are using who would like to learn more we recommend the following introductory texts:

Hair, J. F., Anderson, R. E., Tatham, R. L. and Black, W. C. (1998) *Multivariate Data Analysis*. New Jersey: Prentice Hall.

Lehmann, D. R., Gupta, S. and Steckel, J. H. (1997) *Marketing Research*. New York: Addison-Wesley.

The structure of Part IV is as follows:

1. First we will present the distribution of Template-based ideas in successful new products. It turns out that 70% of successful new products in a variety of categories pertain to one or other of the Templates' structures.

2. Then we present the experiments, which indicated that individuals trained in the Template approach were able to elicit ideas that were superior to those elicited by individuals in various control groups.

3. We conclude this part by demonstrating that Templates are not likely to be found in product failures. These findings lead to the recommendation that Templates be used as a preliminary screening method.

11 Demarcating the Creativity Templates

Mapping research: toward a product-based framework for Templates definition

We initially identified Creativity Templates through a backward analysis of product innovations, a process we call mapping research. We traced the history of a product through its former versions. By portraying the configuration of each product version and, subsequently, examining the stepwise changes between versions, we were able to observe common patterns of change that we later classified into the Creativity Templates. In other words, we tried to discover "laws of product evolution" through the observation of changes in successful product configurations over time. The application of these "laws" to an existing product and predicting a new product thereof is the basis of the Creativity Template approach.

In order to map the configurations of the products, we drew only on their internal dynamics. Our maps consist of product characteristics and reciprocal influences of characteristics. The taxonomy of templates is partially based on and is consistent with the works of Altschuller [3,4], Maimon and Horowitz [5]. The taxonomy and details of doing this are explained below.

As an illustration we present here one of the first mapping research examinations that was performed on soap-related products. The choice of the soap category for analysis was guided by several considerations: (1) large drugstores offer a very large selection of soap-related products; (2) the description and analysis of the various product versions in terms of product configurations and operators of change (according to the Creativity Templates approach) are comprehensive and not exceedingly complex; (3) there is plenty of data available about the historical development and branching out of this category; and (4) soap products are widely used and have fairly well-defined category boundaries (i.e., there is a general consensus as to which product versions should be classified as soap products).

Like most products, soaps have diffused, branched out and multiplied over centuries to satisfy market needs. This branching out process has led to a proliferation of versions and brands in soap products, and to the introduction of several new product categories – soap, shampoo, shaving foams, washing powders, household cleansers and dishwashing detergents.

Over 1,500 product labels of soap-related brands were counted in two large drugstores. Both stores belonged to drugstore chains, which together captured 25 percent of the market share in a city of about 500,000 residents. Figure 11.1 illustrates a partial evolution tree and branching out of the soap-product classes into the first three "generations" of sub-categories.

Shampoo originally emerged from regular bars of soap. The basic idea of the shampoo as a differentiated product manifested itself as liquid soap easily spread through the hair. Two characteristics are relevant here in comparing soap bars with shampoo – consistency and area of application on the body. In the case of the original bar of soap, these two characteristics are not related. The bar of soap has the same consistency regardless of the area of the body over which it is used. The idea of shampoo generates a dependency between these two characteristics: one consistency (solid) for the body and a different consistency (liquid) for the hair. The specific type of relationship between the two characteristics is a step-function.

The commonalties identified in the transition between products were considered to be candidate Creativity Templates. Two experts independently classified the soap-related products according to the Creativity Templates, achieving a 97 percent rate of agreement between them. As each sub-category in Figure 11.2 is considered to be a successful new product, a high proportion of Creativity Templates (78 percent) was arrived at in the development of new products.

This exact mapping procedure was then replicated for three other product classes – hygiene products, bank accounts and sneakers. No new Templates were discovered in these replications.

The mapping study involved rigorous definitions of Templates.

Can Templates explain and predict the emergence of blockbuster products?

The answer to this question is: definitely in the case of some of them, but not all of them, partly because the paradigm we present adheres to rules of parsimony and, as in the case of other comparable explanatory approaches, it does

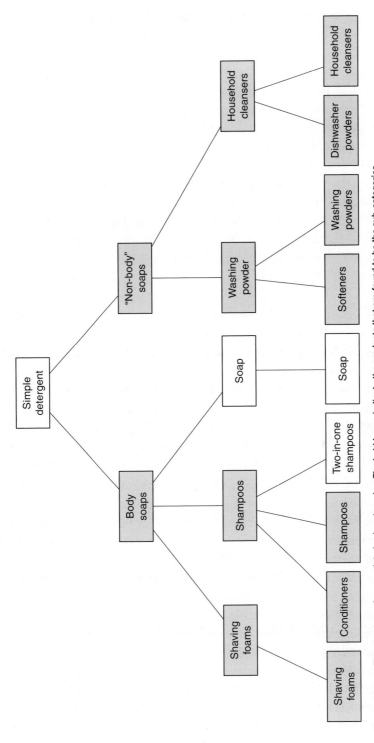

Figure 11.1 The structure of soap-related sub-categories. The bold boxes indicate the products that are found to be the sub-categories.

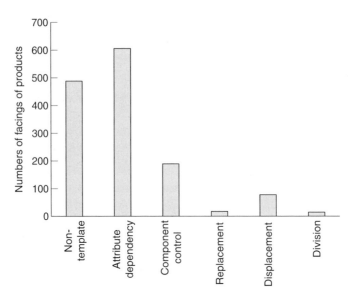

Figure 11.2   The distribution of Templates in soap product facings.

not purport to account for all random events. However, we claim that instead of competing for an exhaustive explanation, the relevant question is whether such new blockbuster products can be uniquely explained by applying the taxonomy of Templates based on past successful product trends.

Consider, for example, 3M's famous Post-It Notes® [see detailed description in 6] which was invented incidentally. Internal testing within the 3M organization produced extensive enthusiastic employee reactions. However, although the results obtained in test marketing conducted in four cities were acceptable, they were not outstandingly optimistic. A careful inspection of the data by the firm's executives indicated that there was, though, a high sales potential when promotion involved free sampling and demonstration. It was that combination of selling effort that elevated purchase intention scores markedly, making Post-It Notes® the most successful new product of 3M by 1984, and subsequently placing it among the top four office supply products, along with copier paper, file folders and sticky tape.

As another example, consider the invention and early introduction of the Walkman® [for a detailed description see 7]. In this case too, the invention was not intended initially. Indeed, even during the first stages following its introduction, the marketers did not envision the success potential of this product. The earlier monophonic Pressman®, which had a recording device, failed and was abandoned. Attempts to make it smaller in size failed too, because the

Table 11.1 Blockbuster products and Template structure

Affiliation to Template	Known new product concept	Formal representation
Replacement	Using the cellular phone billing system instead of the service billing system (e.g., parking – the driver calls to a specified number in order to park and pays via the cellular billing system)	
Displacement	Walkman® (Sony) The recording system was eliminated, enabling a compact size and portability.	
Component control	Post-It Notes® A new link was generated between the table and the note	
Division	The emergence of hair conditioner category from an initially integrated product	

recording system did not then fit. Rather than invest any more effort, it was put aside and used by the company's engineers for their own entertainment. Only after the integration of this concept with that of a light headphone, was the Walkman® concept defined.

These are but two examples of a host of classic new product breakthroughs that can be characterized by two important features; first, they match one of the Templates defined and illustrated previously, and second, the input of consumer needs, derived from assessment of current market trends at the invention stage, did not contribute to the accurate appraisal of the products as major long-term marketing achievements. Table 11.1 provides the formal representation of examples of well-known product successes that can be depicted as having a Template structure. Note that the sequence of events underlying their development can be portrayed in a well-defined manner, and thus, although in reality they emerged accidentally, they could have been foreseen. In order to conduct a more systematic examination of Template-matching ideas underlying blockbuster products, the 50 of the 200 published case histories reported in Freeman and Golden [6] were analyzed. Out of the 23 cases identified as random invention 19 were found to be Template-based.

Obviously, in order to create an operational agenda of stages that might replace the accidental nature of such inventions, two conditions must be met. First, the approach has to be trainable, that is, knowledge of the Templates has to be assimilated by workshop participants, and applied by them in the desired context. Second, measures of creativity and performance must have the capacity to differentiate between inferior and superior ideas and allow for an early screening by those who are experts in market knowledge.

Can training in Templates improve creativity and quality of product ideas?

In the following sections we report studies examining the trainability of the Templates and the value of ideas generated by them, and compare them with those generated by training in rival unbounded methods.

Preliminary study: are Templates trainable?

The purpose of the preliminary study was to provide a first step in assessing whether the Template approach is trainable as an ideation method, and whether individuals responsible for new product decisions in their companies can actually generate Template-based ideas following such training. A workshop was designed to train the participants to use the Template taxonomy and to implement it in developing new product ideas. Five senior employees of a mid-size pharmaceutical company participated in the four sessions of the workshop, each lasting about four hours. The participants were trained in all five Templates (Attribute Dependency and the four component Templates), with about equal time being devoted to each Template. In all cases the training and examples used were drawn from contexts and product categories different from those used in the study. Once it was indicated that a Template was satisfactorily understood and could be used in ideation, the participants were asked to generate product ideas relating to a specific type of medication. Then, the next Template was presented.

Table 11.2 The distribution of generated ideas

Templates	Pre-study	Study 1	Study 2
Attribute Dependency	16	16	16
Component Control	10	10	9
Replacement	6	—	3
Displacement	2	—	—
Division	—	4	3
Total	34	30	31
Non-Template total	20	14	28

In the ideation task, participants were encouraged to use conventional ideation methods involving "total freedom" idea generation, such as in brainstorming, and to elicit as many different ideas as possible. All were previously trained in applying such conventional methods. In addition, idea generation was conducted using the Creativity Template taxonomy. Although we recognized that ideation might have been skewed toward Template usage rather than reliance on conventional methods, the objective of the preliminary study was, nevertheless, to assess whether participants could generate a meaningful number of ideas that match the Templates.

Results and discussion   The results are presented in Table 11.2 (column 1). Altogether, 54 ideas were obtained by the group. The Template procedures yielded 34 ideas (63% of the total number of ideas). Since the main task of the group was to generate ideas for new products, the number of Template-based ideas implies that thinking within bounded scopes is as natural as divergent, unbounded thinking. It should be noted that this workshop represented a "real-life" task for the company: neither the participants nor the instructor were told that the results of their work were also part of an experiment.

How effective are the Templates?

Study 1

This study was field research focusing on the effectiveness of the Creativity Templates. Specifically, it was designed to examine whether the utilization of Creativity Templates in the ideation process leads to superior outcomes.

Procedure   Five senior employees of a credit company participated in a training workshop that consisted of six sessions, each lasting about four hours. The workshop was designed to teach the participants about the Template taxonomy and to train them to implement it in developing new product ideas.

Participants were first trained in the use of conventional ideation methods involving "total freedom" in idea generation, such as brainstorming, synectics, lateral thinking and free association. Training ceased when participants mastered and were able to utilize the methods in ideation. They were then asked to generate as many different ideas as possible. This phase yielded 14 ideas. The next stage was training in the use of Templates, after which participants re-engaged in idea generation, this time based exclusively on Templates. That session yielded 30 ideas. Table 11.2 (column 2) displays the breakdown of ideas into Template and non-Template categories.

About a month after the workshop ended, all the ideas were rated for their overall quality. Eight judges performed this procedure – five participants in the workshop and three senior marketing professionals who had not participated in the workshop. The delay of one month before the rating procedure was designed to counter undesired effects attributable to recall of the classification of ideas as originating from either conventional methods or from Creativity Template procedures. The potential confounding effect of self- versus external-rating was assessed by individual level analysis for each judge, as reported below.

Results and discussion   All eight judges were asked to rate the ideas on a five-point overall quality scale. The ratings were averaged, following an intra-classification procedure, which yielded a reliability level of 0.70. The analysis focused on differences between the ideas derived from the Template procedure and those generated using conventional methods. Comparison between the ratings of the Template-based ideas and those generated using conventional methods indicated an advantage of the former methods ($F(3,40) = 111.37$, $p < 0.0001$). Comparisons between each of the Templates individually and the conventional methods produced significant differences ($p < 0.001$ in all comparisons), whereas no differences were found in comparing the Templates among themselves (in all comparisons, $p > 0.10$ or above).

In order to assess potential bias due to five of the judges being former participants of the workshop, the ratings of the participant judges were compared with those of the non-participants. The mean judgment of the former group was 3.13, whereas that of the latter group was 3.18, with no significant difference between the two ($t(42)<1$, n.s.). The ratings were also analyzed individually for each of the eight judges. In all eight cases, the ratings of the Creativity Template-based ideas were significantly higher than those for ideas generated by the conventional methods (all t's ranged between 2.09 and 4.77 and were significant at least at $p<0.05$ level). Thus, all the judges (participants and non-participants) rated the Template-based ideas as superior.

Study 1 compared Template-based ideas with ideas generated by conventional methods. Although, for the sake of testing the effectiveness of the Template method it was useful to compare it with other methods, Study 1 was confounded by the difference in the timing of the measurement between "total freedom" methods and the restricted thinking approach. In Study 2, the comparison was conducted only after participants had been exposed to both the conventional and the Template approaches. The difference between Template and non-Template ideas depended on whether the ideas fitted the Templates. It should be noted that the objective was to obtain a real life application without preference to the source of the idea – whether it stemmed from the Template approach or from other approaches.

Study 2

Procedure   The procedure in Study 2 was essentially similar to that of Study 1, with two main differences. First, the workshop took place in a bank and involved the domain of saving accounts, with the participants being senior employees responsible for designing and planning of saving accounts. Second, ideas were generated upon completion of the workshop, after balanced training in both "total freedom" and Template-based ideation.

Overall, 59 ideas were generated, 31 of which fitted the Templates and 28 did not. Table 11.2 (column 3) displays the breakdown of ideas into the Template and non-Template categories.

The ideas were subsequently presented for evaluation to two banking experts. These judges were unfamiliar with the methods and were blind to the categorization of ideas into Template-based and non-Template-based. The ideas were rated on the same scale as in Study 1. The mean judgment obtained for Template-based ideas was higher (mean $= 3.91$) than for non-Template-based ideas (mean $= 2.85$, $F(4,54) = 11.24$, $p<0.001$). There was no significant difference among the Template-based ideas (in all comparisons, $p>0.35$ or higher). The breakdown of judgments by the individual Template categories is presented in Table 11.3.

Table 11.3 Mean ratings of the ideas

Template	Study 1		Study 2	
	Mean	Standard deviation	Mean	Standard deviation
Attribute Dependency	3.4700	0.5273	4.0000	0.7071
Component Control	3.6760	0.3257	3.7778	0.7949
Replacement	—	—	3.6667	1.0408
Displacement	—	—	—	—
Division	3.1900	0.8040	4.1667	0.7638
Non-Template	2.3679	0.4717	2.8571	0.8909

How effective is the Attribute Dependency Template?

A further two experiments, reported below, were conducted to examine whether training in the Attribute Dependency Template improves outcomes compared to two widely used unstructured methods – lateral thinking and random stimulation (Study 3) and to a structured method – HIT[1] (Study 4). The focus on this particular Template is important because of its dominance in new product emergence.

Study 3

Overview   The experiment was composed of two phases. In the first phase the idea-generation setting was divided between groups. Participants were trained in the Template approach, or trained in lateral thinking, or trained in random stimulation. Following training, they were asked to generate ideas either regarding baby ointments or mattresses. In the second phase the ideas were rated by senior marketing professionals. All subjects participating in idea generation were university graduates or close to graduation. Their age range was 23–40 with a mean age of 30. Preliminary analysis showed that the groups did not significantly differ in age, education or occupation.

Design and procedure   The 120 subjects were randomly assigned into six groups generated by crossing the training factor (Template, lateral thinking, random stimulation) with the product-category factor (baby ointments,

---

[1] HIT (heuristic ideation technique) is a version of morphological analysis that is designed to manage a thought process which links various marketing concepts in order to form a new combination.

mattresses). Comprehension of the techniques was verified and explanation ceased only when the participants indicated that they could implement them in ideation. The Template training involved about four hours including practice tasks. Training time in the competing methods was about two hours including practice tasks.

The efficacy of the Template approach was assessed by comparing outcomes of the Template training groups (hereafter the TT groups) with those of the lateral thinking training (TL) and random stimulation training (TR) groups. The time allowed for generating ideas (30 minutes) was identical in all six groups. Overall, the six groups generated 277 ideas, 132 for baby ointment and 145 for mattresses.

The ideas generated by the TT group for the ointment category included: an ointment that gives a specific odor upon urination by the baby; a series of ointments that differ in their concentration of active ingredients depending on the sensitivity of the baby's skin; and the introduction of two types of ointment – one for day use and one for night use (higher consistency for increased protection at night by isolating the skin from the urine and lower consistency to allow for skin breathing during the day). Among the ideas generated by the TL group were: an ointment made of natural ingredients and an ointment with scent. The TR group's ideas included long-lasting ointment and a colored (replacing the traditional white) ointment.

Three senior marketing professionals were invited to participate as judges in the evaluation procedure. All three professionals held MBAs, had a record of at least 10 years of experience in the marketing of consumer goods and held high-ranking marketing positions (at least vice president or equivalent level). These professionals were asked to rate the ideas on two scales, chosen in accordance with Finke's [8] suggestion that ideas be assessed by their originality and practical value. These scales are also compatible with the originality and usefulness measures recently adopted in the context of new product design. Accordingly, one scale measured originality with scale anchors (1) "not original at all" and (7) "very original." The second scale measured the overall value of the ideas. The judges were asked to indicate whether they would recommend investment in implementing the idea on a scale ranging from (1) "not recommend at all" to (7) "highly recommend." The judges were blind to the identity of the group members, to one another, to the notion of Templates and to the purpose of the experiment.

Results   Table 11.4 displays the mean originality and value ratings for the three training groups. In the ointment category significant differences were

Table 11.4 Study 3: Mean originality and value ratings

	Originality			Value		
	TT	TL	TR	TT	TL	TR
Raw ideas						
Ointment	5.47^a (0.31)	3.57 (0.82)	3.36 (0.93)	5.97^a (0.43)	3.09 (0.93)	2.82 (0.97)
Mattress	5.83^a (0.50)	3.47 (1.19)	3.22 (0.51)	5.79^a (0.42)	2.71 (0.75)	2.64 (0.83)
Best ideas						
Ointment	5.47^a (0.31)	4.27 (0.45)	4.22 (0.55)	5.97^a (0.43)	3.75 (0.76)	3.68 (0.83)
Mattress	5.83^a (0.50)	4.86 (0.55)	4.66 (0.51)	5.79^a (0.42)	3.27 (0.55)	3.21 (0.67)

Notes:
The numbers in brackets indicate the standard deviation.
^{a} Represents significance (at least at $p<0.05$ level) in the contrast between TT and the combined TL and TR groups.

obtained between the three groups for both the originality of the ideas ($F(2,129) = 52.13$, $p<0.0001$) and the value ratings ($F(2,129) = 911.46$, $p< 0.0001$).

Comparison between the TT group and the TL and TR groups combined indicated that ideas produced by the TT group were superior both in original- ity ($t(129) = 11.6$, $p<0.001$) and in value ($t(129) = 13.85$, $p<0.001$). Comparison between the TL and TR groups showed no significant differences in terms of originality ($t(129) = 1.34$, $p>0.15$) and value ($t(129) = 1.25$, $p> 0.20$).

In planning the analysis it was recognized that unstructured techniques are geared to the production of a large number of ideas varying widely in quality and value whereas structured methods are prescreened and more focused [9]. Accordingly, the small number of ideas produced by the TT group was selected and compared with a matching number of the highest-ranking ideas obtained in the TL and TR groups. The ideas were sorted in descending order of rating. Based on this procedure the top 20 ideas were selected from each group. The second panel of Table 11.4 summarizes the results of this analysis. Significant differences were obtained among the three groups both when orig- inality was the dependent measure ($F(2,57) = 50.50$, $p<0.0001$) and when value served as the dependent measure ($F(2,57) = 69.64$, $p<0.0001$). The comparison between the TT group and groups TL and TR indicated the superiority of the TT group ideas both in originality ($t(57) = 11.04$, $p<0.001$) and in value ($t(57) = 11.79$, $p<0.001$). No significant difference was obtained between TL and TR in either originality ($t(57) < 1$, n.s.) or in value ($t(57) < 1$, n.s.).

This pattern of results was repeated for the mattress category (Table 11.4). As in the ointment task, the analysis first focused on the raw ideas. Significant differences were observed among the three groups both when originality and when value were the dependent measures ($F(2,142) = 45.94$, $p < 0.0001$ for originality, and $F(2,142) = 1411.82$, $p < 0.0001$ for value). Similarly, dominance of the TT group over the TL and TR groups was obtained both for originality ($t(142) = 9.48$, $p < 0.001$) and for value ($t(142) = 111.17$, $p < 0.001$). There was no significant difference between the two non-Template training techniques ($t(142) = 1.27$, $p > 0.20$ for originality and $t(142) < 1$, n.s. for value).

The procedure for comparing the best ideas was subsequently performed resulting in significant differences for the originality ratings ($F(2,57) = 28.61$, $p < 0.0001$) and for the value ratings ($F(2,57) = 140.18$, $p < 0.0001$). Similarly, dominance of the TT group over the TL and TR groups was obtained both for originality ($t(57) = 11.47$, $p < 0.001$) and value ($t(57) = 111.54$, $p < 0.001$). However, between the lateral thinking and random stimulation ideas there was no significant difference ($t(57) = 1.18$, $p > 0.20$ for originality and $t(57) < 1$, n.s. for value)[2].

In addition to the advantage of the Template approach as manifested both in the analysis of the total set of ideas and the "best ideas" note that the highest ranking idea was an outcome of the Template training in both the ointment and the mattress categories.

The comparison between Template training and training in competing unstructured techniques demonstrates the added value of incorporating the Template approach in ideation. It should be noted that a major component of the costs involved in utilizing the TT technique is the relatively longer training time required. However, the high ranking of the ideas generated by the TT group suggests a "self-screening" effect. Moreover, the fact that the TT group generated fewer ideas suggests that the next step of new product ideation (screening followed by concept testing) should require reduced resources and time.

The effect due to training technique in Study 3 may be confounded by the fact that the training time required differs between the techniques and this may

---

[2] Examination of the ideas generated by all three groups in both these experiments indicated that Template-matched ideas were used almost exclusively by the Template trained individuals. Because the division between the Template training and the remaining groups also dichotomizes the use or non-use of the Templates, it can be concluded that Templates are learnable and that they lead to more productive ideation.

account for the results. In addition, Study 3 did not use a group of individuals who were not trained at all prior to idea generation. Such a group might serve as a benchmark for assessing the added value of training in Templates.

Study 4 was devised to address these issues. The technique taught to the control group in Study 4 was HIT (Tauber [10], presented above), which requires equivalent training time and is deemed to be the closest "rival" technique to the Template approach. In addition, in Study 4 the Template approach was compared with a "no training" (hereafter NT) group.

Study 4

Overview   Study 4 used a different product category – drinking glasses. Like Study 3, this experiment was composed of two phases. In the first phase the idea generation setting was manipulated between groups by training in the TT approach, or training in the HIT technique, or not training at all. In the second phase, the ideas were rated by three senior marketing professionals.

Design and procedure   One group consisted of 18 individuals who were specifically trained to use the Attribute Dependency Template (the TT group). The second group was composed of 19 individuals who were trained to apply the HIT procedure (the HIT group). The training in the TT and HIT groups lasted four hours, including practice tasks. The third group consisted of 18 individuals who received no training at all (the NT group). As in Study 3, the efficacy of the Template approach was assessed by comparing outcomes of the TT group with those of the HIT and NT groups. The time allowed for generating ideas (30 minutes) was the same in all three groups. Overall, the three groups generated a total of 82 ideas.

The ideas generated by applying the Attribute Dependency Template included: a drinking glass for babies with color varying according to the temperature of the milk (when the milk is ready for drinking the color of the glass changes from red to blue); a tea cup with varying insulation capabilities according to the temperature of the tea (when the tea is too hot the cup allows fast cooling and, upon reaching an optimal drinking temperature, the cup maintains the desired heat for an extended period of time). Among the outcome ideas generated by applying HIT were: a glass that is purchased with a ready mixture of coffee and milk; a glass that automatically cools fluid when it is poured in; and a glass that glows in the dark for giving children drinks at night. The NT group generated ideas that included glasses that have exotic shapes and a unique combination of colors for a set.

Table 11.5 Study 4: Mean originality and value ratings

	Originality			Value		
	TT	HIT	NT	TT	HIT	NT
Raw ideas						
Glass	5.20^a (1.02)	3.83 (1.35)	4.15 (1.15)	4.82^a (0.90)	2.99 (0.99)	2.91 (0.36)
Best ideas						
Glass	5.20^a (1.02)	3.94 (1.48)	4.15 (1.15)	4.82^a (0.90)	3.59^b (0.65)	2.91 (0.36)

Notes:
The numbers in brackets indicate the standard deviation.
[a] Represents significance (at least at $p<0.05$ level) in the contrast between TT and the combined HIT and NT groups.
[b] Represents significance (at least at $p<0.05$ level) in the contrast between HIT and NT groups.

Three senior marketing professionals were invited to participate as judges in the evaluation procedure. Two of the judges were owners of marketing consulting agencies and one was a senior product manager. All three judges (who did not participate in Study 3) had at least 7 years of experience in marketing positions. The evaluation procedure and scales were identical to those used in Study 3.

Results In analyzing the total set of ideas a significant difference was obtained between the three groups for both the originality of ideas ($F(2,79) = 11.5$, $p<0.001$) and their value ratings ($F(2,79) = 29.8$, $p<0.0001$). The results are presented in Table 11.5.

Comparison between the TT group and the combination of groups HIT and NT indicated that the TT group produced superior ideas both in originality ($t(79) = 3.3$, $p<0.005$) and in value ($t(79) = 11.6$, $p<0.001$). A comparison between the HIT and NT groups showed no advantage in terms of originality ($t(79) = 1.1$, n.s.) or value ($t(79)<1$, n.s.).

Since Tauber's HIT [10] does not confine the number of ideas, the more stringent analysis performed in Study 3 focusing only on the better ideas was repeated in Study 4. Accordingly, the ideas were sorted in descending order of rating and the top 17 ideas were selected from each group, matching the lowest number of ideas generated by the TT and NT groups. The results are presented in the second panel of Table 11.5. Significant differences were obtained among the three groups for originality ($F(2,48) = 5.44$, $p<0.001$) and value

$(F(2,48)=32.3, p<0.0001)$. The comparison between the TT group and groups HIT and NT combined indicated that the TT group ideas dominated both in originality $(t(48)=3.2, p<0.005)$ and in value $(t(48)=11.6, p<0.001)$. HIT training outperformed no training at all in value $(t(48)=2.78, p<0.01)$ although not in originality $(t(48)=(t(48)<1, \text{n.s.})$. In addition to the observed superiority of the TT approach both in the analysis of the total set of ideas and the "best ideas," the highest ranking idea was also Template-based.

Conclusions

The main conclusion of these four studies is that a restricted approach, which draws on past successful Templates of product development, can contribute to the reduction of randomness characterizing the invention of many new and successful products. In addition, such an approach can be utilized to ideate promising product concepts for the future. We have outlined the structure of the Templates by formally specifying the sequence of operators that jointly compose a set of well-defined manipulation processes of product components. The operation focuses on the components that can be uniquely and comprehensively defined and have the capacity to serve as a suitable source for a common language interaction between R&D and marketing personnel in the ideation stage.

The indication that past product-based trends may serve as a useful source for ideation should shift attention from sole reliance on unrestricted ideation approaches. In addition, the Template approach may add value to the ideation methods that draw new ideas from current market needs. While responsive approaches based on analysis of current market needs (obtained by the majority of the market research methods) are highly useful in the more advanced stages of the product life cycle, it is questionable whether these may be similarly powerful in invention and should therefore dominate the ideation process. Reservations about the use of consumer reactions in ideation have been echoed in a number of past studies [e.g., 11]. Urban and von Hippel [12] noted, for example, that while customers may be able to express their opinion about existing products and even predict whether or not they will succeed, they are unlikely to be able to supply researchers with information about products that may be needed in the future.

Competition also does not always serve as a productive source for ideation. Even if attempts to create a competitive edge require knowledge of the entire spectrum of available products, as well as other previously contemplated concepts, the ability of ideation based on that knowledge is questionable. Chan and Mauborgne [13] remarked that the only way to avoid war between firms is differentiation. However, they do not suggest how to do this. They know that they have to be creative, but "thinking out of the box" is only a general strategy. They do not know where to start and what exactly they should be doing.

The recurring Templates in product-based trends serve to prescribe the restricted ideation approach. Recent studies also provide justification for the use of structured, procedural approaches, in that they avoid the "convenient light" syndrome [14] associated with traditional unrestricted ideation methods.

In addition, reliance on random events (or luck) is both inefficient and, for the most part, unreliable. Many new products have emerged as a result of coincidence. However, we do not have reports on the numerous other cases in which random events led to failure or to no ideas at all. The pure assumption of randomness leads to the conclusion that reliance on the occurrence of such an event is not efficient.

Another major conclusion is that the Template approach should be assessed not merely as an ideation method but in its role within the broader scope of integration between R&D and marketing activities. Product-based trends underlying the Templates represent long-term reflections of past marketing realities. The Template approach projects their extension to the present marketing environment. In other words, the product ideas that are the most natural extension of past trends are offered as potential ideas for the present. The main task is, then, the examination of their appropriateness in the current market situation [which is a well-defined problem, see 10]. If the transition of latent need into an existing need resembles an occurrence in the past, it is likely that the Template-matching idea will be recognized as superior in the current market situation as well. This process coordinates and unifies the ideation objectives of R&D and marketing personnel through a comprehensive and common language. Indeed, this convergence represents the ultimate objective of the Template approach.

REFERENCES

1. Goldenberg, J., Mazursky, D. and Solomon, S. (1999) "Toward identifying the inventive templates of new products: a channeled ideation approach," *Journal of Marketing Research*, **36** (May), 200–210.

2. Goldenberg, J., Lehmann, R. D. and Mazursky, D. (2001) "The idea itself and the circumstances of its emergence as predictors of new product success," *Management Science*, 47(1), 69–84.

3. Altschuller, G. S. (1985) *Creativity as an Exact Science*. New York: Gordon and Breach.

4. Altschuller, G. S. (1986) *To Find an Idea: Introduction to the Theory of Solving Problems of Inventions*. Novosibirsk, USSR: Nauka.

5. Maimon, O. and Horowitz, R. (1999) "Sufficient condition for inventive ideas in engineering," *IEEE Transactions, Man and Cybernetics*, **29** (3), 349–361.

6. Freeman, C. and Golden, B. (1997) *Why Didn't I Think of That?* New York: John Wiley and Sons.

7. Mingo, J. (1994) *How the Cadillac Got Its Fins*. New York: HarperCollins Books.

8. Finke, R. A. (1990) *Creative Imagery: Discoveries and Inventions in Visualization*. Hillsdale, NJ: Erlbaum

9. Perkins, D. N. (1981) *The Mind's Best Work*. Cambridge, MA: Harvard University Press.

10. Tauber, E. M. (1972) "HIT: Heuristic Ideation Technique – a systematic procedure for new product search," *Journal of Marketing*, **36**, 58–61.

11. Davis, E. R. (1996) "Market analysis and segmentation issues for new consumer products," *The PDMA Handbook of New Product Development*. New York: John Wiley and Sons, p. 39.

12. Urban, G. L. and von Hippel, E. (1988) "Lead user analysis for the development of new industrial products," *Management Science*, **34** (5), 569–582.

13. Chan, K. W. and Mauborgne, R. (1999) "Creating New Market Space," *Harvard Business Review*, 77, 83.

14. Zaltman, G. and Higie, R. A. (1993) "Seeing the voice of the customer: the Zaltman metaphor elicitation technique," MSI working paper, Report No. 93–114.

12 The Primacy of Templates in Success and Failure of Products[1]

Introduction

In the previous chapter we assessed the value of Templates in terms of their creativity and ideation. The results indicate that templates are likely to enhance creativity. However in marketing and new product development fields it is important to explore the relevance of a method not only with respect to its originality, but also to market success and failure rates. This chapter provides some information about and insights into this applied perspective.

Particularly in view of the distressingly low rate of success in new product introduction, it is important to identify predictive guidelines early in the new product development process so that better choices can be made and unnecessary costs avoided. In this chapter, we posit that the Templates and the Function Follows Forms principle can be utilized as a framework for early analysis based on the success potential embodied in the product idea itself and the circumstances of its emergence. We suggest that these factors, along with already known factors relating to success or failure, may aid estimation of the potential of a concept early in its development.

Predicting new product success

Introduction of new products is a major activity of firms. However, most of the 25,000 products introduced each year in the US fail [1,2]. Because the greatest monetary loss for failed products occurs at the market introduction

---

[1] This chapter is based on joint research work done in collaboration with Don Lehmann from Columbia University.

stage [3], it is critical to gauge reception before introduction and to only continue to promote products that have a high potential for success. Indeed, in view of the fact that expenditures for developing a new product increase as the process advances toward the launch, it is clearly critical for firms to screen out concepts and ideas that are likely to be failures as early in the process as possible [4].

Previous research on new product performance has shown that a wide variety of factors influence the outcome of new product development activities [see, for example, 5 , 6, 7, 8, 9]. These determinants usually involve some combination of strategy, development process, organizational, environmental and market factors.

Research in this area is found in several disciplines including marketing, organizational behavior, engineering and operations management [5, 6, 10, 11]. Much of this research focuses on dyadic comparisons between project successes and failures in an effort to discover the principal discriminating factors, and to provide strategies to enhance success and avoid failure. According to Griffin and Page [12], at least 61 research studies resulting in 77 articles were published on the subject prior to 1993.

Cooper [13] postulated that the success of new product ventures is determined by environmental factors, related to the setting in which a new product is developed, and controllable factors, related to the characteristics of new product activities employed by firms. Since that time, a considerable body of research has been reported. A major comparative study, termed "Project Sappho" [6, 14], examined successes and failures in the area of industrial innovation. In its final form, the project included a total of 43 pairs of success–failure and, by a "pair comparison" technique, factors that discriminated success from failure were identified. Dominant factors were: (1) understanding of user needs; (2) attention to marketing; (3) efficient development work; (4) use of outside advice and technology; and (5) seniority of innovators in their organization.

More recently, Montoya-Weiss and Calantone [5] introduced a taxonomy identifying a logical grouping of the reported measures into those appropriate at *the firm or project level.* In their meta-analysis, they reduced the determinants of product success to 18 factors (product advantage, marketing synergy, technological synergy, etc.). This taxonomy, and variations on it, are widely used in success–failure research [9, 13, 15, 16, etc.]. Focusing on the factors that influence profitability, Ulrich and Eppinger [17] present five dimensions: product quality, product cost, development time, development cost and development capability. Importantly, actual success of a given product is clearly conditional upon the market inclination to adopt it. Market rejection transforms a successful design into a product failure. However, despite all the research that has been conducted in this area, it is still difficult for a firm to determine whether in fact a new product will be successful [12].

Furthermore, a careful review of the literature reveals that little attention has been paid to the contribution to success/failure of the *idea itself* and the unique product configuration implied by the idea. In research to date, the impact of the idea itself has been implicitly included in more complex determinants related to the R&D process rather than investigated directly [6]. For example, Holak and Lehmann [18] use Rogers' [19] typology (relative advantage, compatibility, complexity, communicability and divisibility) to predict product success.

In this chapter, we examine a more extensive classification of the determinants of product success/failure by including a contribution of the *idea* itself (whether it can be classified as a Template-based structure) and the *circumstances of its emergence* (whether it was born as a result of constraints that will be specified shortly) as predictive of actual market results. More specifically, determinants of success and failure used to predict market success are classified into three fundamental groups: (1) *early determinants* consisting of Templates classification and the circumstances of its emergence, (2) *project-level determinants* based on examining the compatibility of the project and the firm (including the execution process), and (3) *market determinants* consisting of market-based knowledge (requiring market research and tests).

The third group is assumed to provide the most accurate forecast because it relies on relevant information about consumer preferences and needs. The determinants of the second group allow for evaluation of the project based on internal observation of the firm's characteristics, which is less costly than engaging in market research. The determinants of the first group allow for evaluation of the product idea at the earliest stage of all – the stage of conception. Shortly we will present an assessment of the power of the first group of determinants to predict product success. If successful in demonstrating the predictive power of the Templates taxonomy, the results can be used either to channel the ideation process into those types of ideas that have a higher probability of success ("self screening" ideation) or as an early screen for the likelihood of success.

Early determinants

Templates of product change

The theory presented in previous chapters of this book questioned whether new products should evolve solely on the basis of knowledge derived from *market-based* information, or whether there is an intrinsic *product-based* scheme underlying the evolution of successful products. Our main thesis is that the Templates are identifiable, objectively verifiable and generally observable, and that these *Templates* can facilitate productive and focused ideation. Because Template-matched ideas were evaluated as more effective than ideas that are not affiliated with Templates, we expect ideas that can be ascribed to Templates to be more successful.

Source of idea

Although the importance of protocols is noted in Montoya-Weiss and Calantone [5], so far the marketing literature has paid little attention to the way in which an idea is generated as a possible predictor of its success. The literature on ideation and creativity points out, however, that the quality of ideas changes when alternative cognitive processes (and circumstances) are involved. According to Finke, Ward and Smith [20, 21], ideas are composed of functions (e.g., consumer needs) and their relation to forms (e.g., solutions). They identified three types of cognitive search for ideas that may be relevant to new product ideation: (1) Identifying or defining a function and then performing an exploratory search for a suitable form; (2) identifying a form followed by an exploratory search for a meaningful related function; and (3) creating the generalization of a predefined, restricted function–form relation. When none of these exist, the efficiency of the process and the quality of the ideas are reduced dramatically.

Adopting these findings to the new product ideation context leads to a three part classification of source variables: (1) "*Need spotting*" – when need identification precedes product (form) development; (2) "*solution spotting*" – when a form is identified and the inventor searches for a suitable need (use) or both the need and a solution are identified concurrently (usually as an improvisation); and (3) "*mental invention*" – when according to the inventor's report the idea is based on a decision to innovate and on an internal cognitive process rather than on external market stimulus (note here that the decision to innovate ignores the recommended "market attention" approach). To these three classes of source variables (identified via self-reports from the individuals involved in the development process), we add two variables related to marketing, namely, (4) *market research for new products* [see, e.g., 22] – when a need is identified by marketing analysis and a suitable product is then developed and (5) *"following a trend"* [1] – when a product is developed to follow a market trend in a different class of products.

For example, the first bandage was designed by a husband in response to the need to stop his wife's bleeding ("need spotting"). In contrast, Vaseline was invented after a chemistry student identified and admired a unique feature in a certain oil residual, and then searched for a suitable benefit ("solution spotting"). The decision to introduce a solid shampoo was reported as an idea that was suddenly born in the mind of a marketer with no "market attention," and, as such, it can be classified as a "mental invention." Pepsi Clear® is an example of "following a trend" because the "clear" trend existed before Pepsi adopted it and there was no relevance to any fundamental features of Pepsi.

Consistent with Finke et al. [20, 21], it is posited that the effectiveness of an idea increases when limits are set on the scope of explored possibilities. In this case, spotting a need or a solution provides cues for an idea before its actual conception. This is consistent with findings showing that a limited search within a confined set of possibilities has a positive effect on the quality of ideas [21, 23, 24].

The proposition that solution- and need-spotting circumstances lead to superior ideas is supported by a series of studies conducted by Von Hippel [see 10, 25, 26] on "lead users." In his work, lead users were found to possess unique information about future needs. By creating solutions to their own problems, they were frequently able to predict new and successful products and often consumers' improvisations provided the basis for formalization of new products. By understanding their needs and problem solutions, lead users provide useful data in a multi-stage process. According to our notion of idea protocols, "need spotting" and "solution spotting" offer signals for successful ideation. Further, empirical tests have validated the proposition contained in the model and elaborated by Finke et al. [20], termed "Geneplore," which suggests that ideas based on solution spotting are superior to those based on need spotting.

Unlike the cases of "solution spotting" and "need spotting," we expect trend following to have negative effect on market performance. Altering a product according to existing market information (i.e., identifying the trend) seems to have a positive effect on the quality of an idea. However, the relevance of a trend to a product is low in many cases (e.g., a clear Pepsi), partly because the idea generation process consists of an attempt to mimic other ideas rather than to generate novelty.

Project-level determinants of new product success

As already mentioned, previous research has provided determinants that examine the match of a new concept with a firm's resources. These determinants are postulated to play an important role in transformation of an idea into a product. As stated, the aim of this chapter is to evaluate the contribution of early determinants to product success and to develop a unified model that incorporates both the early and the project-level determinants.

In view of the importance of an early estimation of the market response to a product, concept tests have become a widely used tool for "go–no go" decisions. However, these tests have some limitations such as inflated or deflated purchase intention ratings and the short time span allocated to consumer reaction [27]. Another shortcoming of concept testing is that it focuses on needs and purchase intention, ignoring other factors that become evident only when the product is presented in its final configuration and design [4]. Extensive research has been conducted in order to improve the predictability of success vs. failure, and to evaluate success probability earlier. The findings generally indicate that factors related to R&D improve market performance, and that the characteristics of new product development activities can be controlled by firms. Below we elaborate the main project-level determinants that have been found to affect the market performance of new products.

1. *Newness to the market* The introduction of products that are "new to the market" can potentially lead to market share gains [see 28]. Indeed, products may be rejected due to their newness or premature introduction. Innovative products form a significant component of a company's offerings. Booz, Allen and Hamilton [29] classified 700 product introductions according to "newness to the market" and "newness to the firm." Of these, 17% were classified as having high market newness, 10% as having a high company newness, and 7% low company newness. Wind and Mahajan [30] noted the disproportionate effort currently devoted to "me too" products (e.g., line extensions, improvement of current products and cost reduction). "Me too" new products are more than twice as prevalent as "really new products" [29]. To illustrate how newness to the market was classified in the present study, consider the introduction of the first shampoo. The fact that people were already using soap bars to wash their hair (prior to the introduction of shampoo) led shampoo to be classified as a new but not radically (i.e., moderately) new product.

2. *Newness to the firm* Tushman and O'reilly [11] review the interplay between revolutionary and incremental innovation and argue that, in order to survive, it is crucial for firms to implement both. However, in the context of a specific product, newness to the firm may lead to failure. Griffin [31] argued that firms are reluctant to adopt inventions that are not consistent with their current activities. This is in line with Cooper's (1985) finding that newness to the firm[2] is correlated with failure rather than success. For example, the fact that the first shampoo was introduced by a soap manufacturer suggests that it was not very new to the firm.

3. *Changes in technology* Typically, addressing more needs and improving system performance entails changes in the system [32]. The market does not generally respond enthusiastically to large-scale changes. Here, changes in technology are measured in relation to existing technologies in the field, and the changes required in the firm's technology in order to manufacture the product. Sanjav, Dongwook and Dae [16] found that newness of the production process is correlated with product failures. In addition, managers often find it difficult to adjust to new technologies successfully [33, 34]. We classify the level of required technology change into three groups: minor, moderate and major. Based on this classification, the first amphibious car required a major technology change and the first shampoo (which was in fact a liquid soap) required a minor technology change.

In general, newness to the market is expected to have a positive impact on product success whereas newness to the firm and technological change are expected to have a negative effect. In other words, the market favors innovative products that do not require major adjustments to produce them. This market preference can be termed "secure progress" – by rejecting highly complex products, the market exerts pressure on companies to produce new products based on existing resources and technology. The plausibility of this hypothesis is supported by previous findings suggesting that innovation adoption usually occurs more readily in the case of products that appear less complex to the consumer [35, 36].

4. *Product offering (the primary advantage of the product)* The importance of the product offering is discussed extensively in the marketing literature. Based on a review of the literature [e.g., 5, 6] and examination of product introductions, we classify products according to their principal offering into the following six groups: (1) technology-stretching products (superior technology introduced into an existing product); (2) need-addressing products (a new and important need satisfied by the new product); (3)

---

[2] Newness to the firm is often defined in terms of newness of the R&D and newness to the market.

economical products (the purchaser saves money or other resources due to lower price or more economical usage); (4) trend-gimmick products (the product offers a gimmick without any other benefit or mimics a non-relevant trend in a remote product); (5) segment-focused products (a product adapted to better fit a market segment); and (6) formalization products (a product that incorporates existing improvisations or consumers' habits) [see also 10, 25].

A product may fall into one or, albeit rarely, more than one, category in this classification. For example, in this study an amphibious car was classified as a technology-stretching product, whereas shampoo was classified as both a need-addressing product and a formalization product (because people used tiny bars of soap and powders to wash their hair before shampoo was introduced). A solid shampoo was classified as a trend-gimmick product, because of its relatively unimportant declared benefits. The new generation of concentrated washing powders was classified as an economical product and a shampoo for treated hair as a segment-focused product.

Hypotheses regarding the predictive power of Templates and other early determinants

The predictive power of the early analysis variables was tested using the two-class Templates typology and the four-class "idea–source" typology described above.

Templates

> **H1a:** Affiliation to the Attribute Dependency Template has a positive effect on product success.
>
> **H1b:** Affiliation to a component Template has a positive effect on product success.

Source of the idea

> **H2a:** Focus on external cues through need spotting or solution spotting has a positive effect on the success of a product.
>
> **H2b:** Mental ideation has a negative effect on product success.
>
> **H2c:** Products based on following a trend or gimmicks are less likely to be successful.

Early analysis performances

> **H3:** Early analysis determinants (Template, source of idea) have the capacity to contribute to the prediction of market success beyond the project-level determinants.

Table 12.1 Success vs. failure determinants in the data of Study 1

Determinant	Variable	Failure (%)	Success (%)
Template	Attribute[a] (18)	12.1	87.9
	Component[a] (24)	4.2	95.8
	Not a Template[a] (31)	83.9	16.1

Notes:
The numbers in the brackets indicate the sample size.
[a] $p < 0.01$.

Study 1: predicting success of patented products

Method   A set of 70 detailed cases of successful and unsuccessful consumer products was collected (41 successes and 29 failures). In order to avoid fuzziness and inaccuracy in deciding whether a product is or is not a failure [12], we define a failure as: (1) a product that was totally rejected by the market and ceased to exist; or (2) a product that failed in market tests resulting in a decision to abort its introduction. Only products that generated substantial positive financial results were defined as successes. The data was obtained from the Israeli patent office, and the 70 patents were chosen randomly from three different categories: kitchen devices, garden tools and car devices. The inventors of each patent were contacted and interviewed in order to define each patent as a success or failure. To ensure that the rejection of a new product was not temporary, patent applications after 1990 were not included.

Each product was classified according to the Template determinants detailed earlier. The classification process involved submitting the cases to a group of judges trained in Template identification. The research is detailed in Goldenberg, Lehmann and Mazursky [37]. In order to test if the proposed Templates are redundant, the correlation between the Template variables was examined. A logistic regression equation was performed to predict product success based on the Templates.

Results   The average agreement between the judges (alpha) was high (alpha $= 0.89$). The correlation between the two Template variables was 0.22 and, thus, they are not redundant. The distribution of successes and failures, broken down separately for each Template, is summarized in Table 12.1.

The logistic regression analysis indicated that a high proportion (89.5%) of the failures and successes can be predicted by the model (see Table 12.2). Table

Table 12.2 Logistic regression prediction results for Study 1

	Predicted failure	Predicted success	Correct (%)
Observed failure	26	3	89.66
Observed success	5	36	87.80
Overall			88.57

Table 12.3 Coefficients of the logistic regression for Study 1

Variable	B	t	Sig
Constant	-1.64	15.27	0.000
Attribute Dependency	3.52	7.54	0.000
Component Templates	4.65	16.82	0.000

12.3 presents the coefficients and their significance. In general, they match the individual variable results, with the two Template predictors significant at the 0.01 level. Clearly, products that follow the Template structure have a greater likelihood of success.

Study 2: the unified model

The aim of Study 1 was to assess the predictive power of Template variables. The relatively small size of the sample and the inability to control for bias attributable to inventors' self-reports in Study 1 made it difficult to consider other variables. The purpose of Study 2 is to allow us create a unified model of idea evaluation, using template determinants (H1a, b), and source-idea determinants (H2a–c) and to compare their predictive power to that of the project-level determinants described in the existing literature (H3).

Method   A set of 127 detailed cases of successful and unsuccessful consumer products was collected (70 successes and 57 failures). The data came from three different books [1, 38, 39] each containing data on successes and failures of new products. We avoided the fuzziness and inaccuracy of deciding whether a product is or is not a failure [12] by using the same criteria as in Study 1, which is in line with this book's own criteria; only products that generated substantial positive financial results were included in the books as successes. Only four product cases were excluded due to classification ambiguity.

In view of the fact that the authors of the books were blind to the purpose of this study, and the absence of any interaction with them, it was assumed that this sample was free of any bias related to the determinants. To assess the reliability of the reported cases, the details were contrasted with a different source, the *Encyclopedia of Consumer Brands* [40]. A sample of 30 cases was examined; no discrepancies were found between the data sources.

The cases were classified according to the determinants detailed earlier (i.e., Template affiliation, idea-source and project-level determinants). The classification process involved submitting the cases to three different groups of three judges, selected for their area of expertise (i.e., technology level was judged by engineers, newness to the market by experienced marketers, Template identification by trained judges). The judges were considered experts in their field; all had at least eight years experience and held senior positions in firms involved in new product introduction and innovation. The categories used for product classifications are summarized in the appendix to this chapter.

As in Study 1, training each group of judges involved a 30-minute session in which the determinants were explained and demonstrated with examples of existing products. The judges were not exposed to the notion of classification of failure or success. The training quality and judgment requirements were assessed following a similar procedure to that of Study 1.

Inter-judge reliability was tested. In order to test the redundancy of the proposed determinants, a factor analysis was performed. A logistic regression equation was then created to predict product success based on the early determinants (Template affiliation and idea-source), project-level determinants, and a combination of both groups of determinants. To estimate the contribution of the template and idea-source determinants to the prediction of success, a nested-model test was performed. Finally, we removed the effect of potential covariates (e.g., durable vs. consumable products, and industry growth) by including them in the analysis.

Results   The average inter-judge agreement was high in all three groups (average alpha $= 0.89$, 0.92, 0.89). The correlation matrix revealed that the highest correlation between variables was -0.39 (between "newness to the market" and "trend following"). The second highest correlation was only 0.27. Thus, the determinants are not highly correlated. In addition, a factor analysis revealed that there are 11 factors with eigenvalues larger than one, which explain 72.6% of the variance. The eigenvalues decrease without a noticeable elbow, suggesting that no clear groupings exist. Therefore, we use specific determinants rather than factors in subsequent analysis.

The distribution of successes and failures, broken down separately for each determinant variable, is summarized in Table 12.4. Most of the differences are

Table 12.4 Success vs. failure determinants (Study 2)

Early deter- minants	Variable	Failure (%)	Success (%)	Project-level determinants	Variable	Failure (%)	Success (%)
Templates	Attribute[a] (24)	8.3	91.7	Newness to the market	High[a] (39)	43.6	56.4
	Component [a](31)	12.9	87.1		Moderate[b] (87)	30.4	69.6
	Not a Template[a] (76)	67.1	32.9		Low[a] (32)	75	25
Source of idea	Need spotting[a] (34)	35.3	64.7	Newness to the firm	New to the firm[d] (87)	47.5	52.5
	Solution spotting[a] (17)	11.8	88.2	Product offering	Need addressing[a] (54)	24.1	75.9
	Market research[a] (11)	27.3	72.7		Economical[d] (5)	60	40
	Random event[b] (7)	14.3	85.7		Segment- focused[d] (24)	37.5	62.5
	Mental invention[a] (40)	67.5	32.5		Trend, gimmick[b] (39)	59.0	41.0
	Trend following[a] (19)	78.9	21.1		Technological superiority[c] (24)	91.0	9.0
					Formalization[b] (27)	25.9	74.1
				Technology change	Major[a] (24)	79.2	20.8
					Moderate[d] (33)	57.6	42.4
					Minor[a] (69)	26.1	73.9

Notes:
The numbers in the brackets indicate the sample size.
[a] $p<0.01$, [b] $p<0.05$, [c] $p<0.1$, [d] p not significant.

consistent with the hypotheses – low levels of newness to the market, major technology change, following a trend, and mental invention tend to be associated with failure. By contrast, meeting needs, economical products, segment-focused, formalization products, minor technology change, need and solution spotting, and products that match a Template tended to be associated with success.

To further investigate the role of each of the early determinants (Template affiliation and idea-source), a logistic regression analysis was performed. The results indicate that most (81.9%) of the failures and successes can be correctly predicted by the Template and idea-source variables (see Table 12.5). Table 12.6 presents the coefficients and their significance. In general, they match the

Table 12.5 Logistic regression prediction results for the early determinants (Study 2)

	Predicted failure	Predicted success	Correct (%)
Observed failure	49	8	85.96
Observed success	15	55	78.57
Overall			81.89

Table 12.6 Coefficients of the logistic regression

Variable	B	t	Sig	Variable	B	t	Sig
Constant	−0.18	0.05	0.83	Need spotting	−0.26	0.10	0.74
Attribute Template	2.94	12.51	0.000	Random event	0.05	0.00	0.96
Component Templates	2.58	15.72	0.000	Marketing research	0.04	0.00	0.96
Mental invention	−1.50	3.32	0.07				
Trend following	−1.81	2.94	0.08				
Solution spotting	2.04	4.13	0.04				

Table 12.7 Logistic regression prediction results for the project-level determinants

	Predicted failure	Predicted success	Correct (%)
Observed failure	45	12	78.95
Observed success	14	56	80.00
Overall			79.53

results obtained for the individual variables, with three variables significant at the 0.05 level and two at the 0.1 level, having all been identified as relevant by the previous analysis.

To further evaluate the role of the Template affiliation determinant and the idea-source in market success estimations, a nested model was tested. First a logistic regression was applied with the project-level variables as predictors. The results indicate that 79.5% of the failures and successes were correctly predicted (see Table 12.7). This result replicates findings of previous studies in new product successes cited earlier: The same factors were found to govern the success and failure, which provides evidence for the external validity of our sample.

We then applied the logistic regression again, including the proposed early determinants to form a unified model. The percentages increased to 92.1% correct predictions (see Table 12.8), indicating a significant incremental

Table 12.8 Logistic regression prediction results for the unified model (Study 2)

	Predicted failure	Predicted success	Correct (%)
Observed failure	52	5	91.23
Observed success	5	65	92.86
Overall			92.13

Table 12.9 Coefficients of the logistic regression for the unified model (only significant determinant included, all other determinants are not significant)

Early determinants	B	t	Sig	Other determinants	B	t	Sig
Attribute Template	3.66	7.74	0.005	Low market newness	−1.95	5.02	0.02
Component Templates	2.95	10.55	0.001	Meeting needs	1.75	4.39	0.03
Mental invention	−2.27	6.41	0.01	Formalization	3.03	12.53	0.003
Solution spotting	1.83	2.55	0.1	Low technology leap	4.46	16.43	0.000
Trend following	−2.40	3.50	0.06	Constant	−2.87	9.47	0.002

contribution of early determinants in success prediction even in later evaluations, when more project-level information is available. Table 12.9 presents coefficients of the significant variables of the unified model. Comparison of the unified model with the early determinants again suggests that the early analysis alone accounts for a substantial amount of the predictive power of the full set of variables.

Effect of covariates

One possible explanation for the predictive ability of the early analysis determinants is that they reflect the market conditions into which the product is introduced, and that it is these covariates, and not the early or project-level determinants, that govern success or failure. Therefore, we examine some obvious covariates that might explain success. Specifically, four covariates were tested: (1) consumer vs. industrial products, (2) durable vs. consumable products, (3) growth rate in the industry, and (4) high-tech vs. low-tech products. Each of the 197 products (from Study 1 and Study 2) was categorized with respect to these four variables. Initial analysis showed that the proportion of successes vs. failures for each covariate in the sample was not significantly different from 50% (Table 12.10). When logistic regression was performed using the four covariates as predictors, only 59.4% of the successes and failures were correctly predicted, and the coefficients of the covariates were not

Table 12.10 Success vs. failure covariates

Covariate		Failure (%)	Success (%)	p value
Product classification	Consumer (186) vs.	43.0	57.0	0.33
	industrial (11)	54.5	45.5	
Product classification	Consumable (83) vs.	39.5	60.5	0.19
	durable (114)	49.4	50.6	
Growth rate of the industry	High (43) vs.	46.5	53.5	0.73
	low (154)	42.9	57.1	
Industry classification	High-tech (24) vs.	62.5	37.5	0.17
	low-tech (173)	41.0	59.0	

Note:
The numbers in the brackets indicate the sample size.

significant (p values varied from 0.131 to 0.588). When a logistic regression was performed including both the idea variables and the four covariates, the increase in the correct prediction was minimal (1.9%). Further, in this regression none of the covariate coefficients was significant. By contrast, all the coefficients of the variables significant in the unified model remained significant and were similar in magnitude. A loglikelihood test confirmed that the covariates do not add significant predictability to the current regression.

Why early determinants can predict success

Given the high failure rate for new products, it may be surprising to learn that certain factors in general and Templates in particular provide a strong clue as to a product's eventual success or failure. This paradox is resolvable by considering the information processing done by potential adopters both on their own and in response to persuasive communication (e.g., advertising).

First, remember that humans in general avoid rather than seek innovation. Part of this is based on evolution and adaptation which makes individuals comfortable with both the familiar and the fathomable. In other words, radical changes are likely to be rejected and minor ones ignored. This leads to the notion of the optimal or "just right" level of innovation and explains why modest innovations tend to be more successful than trivial or radical ones. In other words, a successful innovation must at the same time be both new and easy to comprehend.

Following Goldenberg, Mazursky and Solomon [41, 42, 43, 44] we argue in this book that the relations between products and their market is evolutionary. When a change in the external environment (e.g., market preferences) occurs, products that do not adapt to the new condition cease to exist. Over time market changes leave traces in product configurations that can be identified as product-based trends. Those trends, crystallized as Templates, provide the skeletons from which numerous new successful product ideas are spawned.

One reason early determinants such as Templates predict success is that they channel information processing into routes that invoke a perception of innovativeness. The fact that certain structures are conceived as superior is consistent with the view of the brain as a self-organizing system where paths of minimum energy serve as attractors for responses or preferences [45]. There is mounting evidence that such schemes facilitate consumer receptiveness to new products. Goldenberg, Mazursky and Solomon [41] identified six groups of Templates in the context of advertising, some of which are similar to the Replacement and the Attribute Dependency Templates described earlier. Ulrich [46] describes function sharing in the context of mechanical design, whereby an object already carrying out one function is assigned another. In the context of new product development function sharing and the Replacement Template are identical. The Templates discovered in the context of new products comply with what Maimon and Horowitz [47] term a *closed world condition* which, in the domain of engineering problem-solving, was shown to be one of two sufficient conditions needed for ideas to be perceived as inventive. Relatedly, Nam Suh [48] offers a set of axioms characterizing good design. Designs that fit the axioms involve minimization of the "information content of the design" (the Information Axiom) or the total number of parts [see also 49]. The best design according to Nam Suh is "a functionally uncoupled design that has the minimum information content" [48, p. 48].

The advantage in processing Template-based ideas is in the type of knowledge transfer generated when consumers try to identify similarities between a new idea and previously known ones. Holland et al. [50] drew a distinction between *surface* similarities (e.g., chemical ingredients) and *structural* similarities between the target (new idea) and the base (existing ones). If the idea is new, comparison based on surface properties may be limited due to the dearth of superficial links between the target and the base. Unlike surface properties, Templates are "deep" structures [51] that rely on regularities which have been proven successful in other contexts (possibly even by the same consumers). Template-based ideas are likely to be perceived as new, yet "familiar." The upshot of this is that Template-based ideas on the one hand create a sense of new superior products. On the other hand, they minimize the risk of rejection associated with their novelty, because they are structurally similar to other successful products.

A second and related reason that certain determinants predict success is discussed in the literature on product diffusion. Rogers [19] points out certain factors which enhance or retard adoption (the complexity of the product, the degree to which it may be tried on a limited basis, etc.). These factors determine the benefits related to using the new product versus the level of perceived risk in use [52]. A related factor influencing adoption or rejection is the communicability of the innovation: According to Rogers [19] an important aspect in the diffusion of innovation is that the innovators and potential users differ in expertise. When a change agent is more technically competent than his or her clients this frequently leads to ineffective communication. As a consequence, the fewer new things to convey, the easier it is to convey them.

Taken together these points suggest that consumers' mental models favor innovations that consist of something new surrounded by familiar product attributes. Since Templates involve changing only one or a few attributes or components, they may be more compatible with consumers' mental models [53]. Communicability of Template-based ideas is enhanced because a high perceived benefit obtained with minimum configuration changes leads to favorable evaluation since it does not undermine one's knowledge structure.

Conclusions of the empirical studies

We found that adoption and rejection can be estimated using early determinants that can be constructed by inspecting the idea itself (Template affiliation) and the circumstances of its emergence (protocol). More precisely, successful products tend to (1) fit one of the Template groups, and (2) involve a solution to a customer problem. In contrast, products developed in isolation by the inventor or products that attempt to mimic a popular trend from other products were generally unsuccessful.

These early determinants allow for the prediction of an idea's market potential in the very first stages of its emergence. It is not argued that market response can be predicted accurately without market research. However, by using the early determinants, ideas can be screened before progressing to the concept level of development and market testing. Further, the incorporation of early determinants into models of new-concept valuation (using project-level determinants) is likely to increase their predictive power even if applied later in the process of product development.

These conclusions are subject, of course, to the limitations of these studies. A sample consisting of products from different categories or new products that differ less dramatically in outcome may lead to weaker or even different results. Further, a host of potential covariates remain to be investigated. We have demonstrated that the early determinants dominate such obvious covariates as consumer vs. industrial products, durable vs. consumable products, growth rate and low-tech vs. high-tech in predictive power. However, other important covariates may exist (e.g., controllable factors of new product management [6, 54]. Another effect that should be considered is that the sample in Study 2 does not include products that failed to reach the market. For example, it is plausible to assume that radically new products (to the market, or from the technology change perspective) are associated with failure (as verified in Study 2). Hence, the results of our study have only been demonstrated for those new products that reach the introduction stage. In addition, it is important to note that the expected value of a developed new product is a function of probability of success and revenue. Our model does not consider the "risky but highly rewarded" projects; the determinants discussed are applicable solely to the probability of success.

Appendix

To clarify our classification procedure, we illustrate below examples of cases and their coding[3].

Table 12.11 Examples of cases and their coding

Case	Description	Template	Source	Project-level determinant
Amphibious car	A car designed to be also used as a boat.	Not a Template	Mental invention	High newness to the market, major technology change, new to the firm, segment focus
Bandage	A husband applied a gauze and adhesive to his wife's injured hands. Johnson & Johnson adopted his "lead user" based concept	Component Control	Solution spotting	Moderately new to the market, minor technology change, not new to the firm, need addressing

---

[3] Note that the source classification is based on the details of the case which are not provided here.

Table 12.11 (*cont.*)

Case	Description	Template	Source	Project-level determinant
Coffee drip filter	The inventor spotted when her friend smiled, specks of black coffee grounds on shiny white teeth	Component Control	Need spotting	High newness to the market, minor technology change, new to the firm, need addressing
Pepsi Clear®	A transparent Cola that mimicked the trend of transparent drinks	Not a Template	Trend following	Not new to the market, minor technology change, not new to the firm, trend/gimmick offering
Vaseline®	A residue from an oil-rig pump (called "rod wax") was reduced to a moist white jelly as part of a laboratory experiment	Displace-ment	Solution spotting	High newness to the market, minor technology change, new to the firm, need addressing
Polavision®	Polaroid camera that produced a 2.5-minute movie	Not a Template	Mental invention	High newness to the market, major high technology change, not new to the firm, need addressing
Thermos® flask	A scientist experimenting with low-temperature liquefaction of gases created a device in which he placed a flask inside another and created a vacuum between them that prevented heat transfer	Attribute Dependency	Solution spotting	High newness to the market, minor technology change, new to the firm, need addressing
Smokeless cigarettes	Cigarettes that burn without smoke	Not a Template	Need spotting and market research	High newness to the market, major technology change, not new to the firm, need addressing
Conditioner	An idea to divide shampoos into two different product categories by excluding the conditioning ingredients and introducing them as a separate product	Division	Market research	Moderately new to the market, minor technology change, not new to the firm, segment focus based offering
Ultrasonic device against cats	A device that produces ultrasonic waves to scare cats from sitting on warm cars.	Not a Template	Mental invention	High newness to the market, moderate technology change, new to the firm, need addressing

Table 12.11 (*cont.*)

Case	Description	Template	Source	Project-level determinant
Snacking doll	A doll designed with a little motor and a set of gears that powered the doll's jaw (which included teeth) allowing it to chew plastic carrots	Not a Template	Mental invention	Moderately new to the market, moderate technology change, not new to the firm, gimmick
Unwaxed candle	A brushed aluminum butane candle was introduced to replace the wax candles	Not a Template	Mental invention	High newness to the market, major technology change, not new to the firm, economical offering

REFERENCES

1. McMath, R. M. and Forbes, T. (1998) *What Were They Thinking?* New York: Times business-Random house.
2. Bobrow, E. E. and Shafer, D. W. (1987) *Pioneering New Products: A Market Survival Guide.* New York: Dow Jones-Irwin.
3. Robertson, T. S. (1971) *Innovative Behavior and Communication.* New York: Holt Rinehart and Winston Inc.
4. Dolan, J. R. (1993) *Managing the New Product Development Process.* Reading, MA: Addison-Wesley.
5. Montoya-Weiss, M. M. and Roger Calantone, R. (1994) "Determinants of new product performance: a review and meta-analysis," *Journal of Product Innovation Management,* 11, 397–417.
6. Freeman, C. (1982) *The Economics of Industrial Innovation.* Cambridge, MA: MIT Press.
7. Virany, B., Tushman, L. M. and Romanelli, E. (1992) "Executive succession and organization outcomes in turbulent environments," *Organization Science,* 3, 72–92.
8. Cooper, G. R. and Kleinschmidt, E. J. (1987) "New products: what separates winners from losers," *Journal of Product Innovation Management,* 4 (3), 169–184.
9. Garry, L. L. and Yoon, E. (1989) "Determinants of new Industrial product performance: a strategic reexamination of the empirical literature," *IEEE Transactions on Engineering Management,* EM-36, 3–10.
10. Von Hippel, E. (1988) *The Source of Innovation.* Oxford: Oxford University Press.
11. Tushman, L. M. and O'reilly, C. A. (1996) "Ambidextrous organizations: managing evolutionary and revolutionary change," *California Management Review, Berkely* 38 (4), 8.
12. Griffin, A. and Page, A. L. (1996) "PDMA success measurement project: recommended measures for product development success and failure," *Journal of Product Innovation Management,* 13, 478–496.
13. Cooper, G. R. (1979) "Identifying industrial new product success: project newProd," *Industrial Marketing Management,* 8 (2), 124–135.

14. Rothwell, R. (1985) "Project Sappho: comparative study of success and failure in industrial innovation," *Information Age*, 7 (4), 215–220.

15. Griffin, A. (1996) "Obtaining information from consumers," in *PDMA Handbook of New Products Development*, Toronto, NY: Wieley, pp. 154–155.

16. Sanjav, M., Dongwook, K. and Dae, H. L. (1996) "Factors affecting new product success: cross-country comparisons," *Journal of Product Innovation Management*, **13P**, 530–550.

17. Ulrich, K. T. and Eppinger, S. D. (1995) *Product Design and Development*. New York: McGraw-Hill.

18. Holak, L. S. and Lehmann, D. R. (1990) "Purchase intentions and the dimensions of innovations: an exploratory model," *Journal of Product Innovation Management*, 7, 59–73.

19. Rogers, E. M. (1995) *Diffusion of Innovation*. New York: The Free Press.

20. Finke, R. A., Ward, T. B. and Smith, S. M. (1992) *Creative Cognition*. Cambridge, MA: MIT Press.

21. Finke, R. A., Ward, T. B. and Smith, S. M. (1995) *The Creative Cognition Approach*. Cambridge, MA: MIT Press.

22. Crawford, C. M. (1977) "Marketing research and new product failure rates," *Journal of Marketing*, **41** (April), 51–61.

23. Perkins, D. N. (1981) *The Mind's Best Work*. Cambridge, MA: Harvard University Press.

24. Weisberg, R. W. (1992) *Creativity Beyond The Myth Of Genius*. New York: W. H. Freeman Company.

25. Von Hippel, E. (1989) "New product ideas from lead users," *Research Technology Management*, **32** (3), 24–28.

26. Urban, G. L. and Von Hippel, E. (1988) "Lead user analysis for the development of new industrial products," *Management Science*, **34** (5), 569–582.

27. Iuso, B. (1975) "Concept testing: an appropriate approach," *Journal of Marketing Research*, **12** May, 228–231.

28. Urban, L. G., Carter, T., Gaskin, S. and Mucha, Z. (1986) "Market share rewards to pioneering brands: an empirical analysis and strategic implications," *Management Science*, **32** (6), 645–659.

29. Booz, Allen and Hamilton (1982) *New Product Management for the 1980's*. New York: Booz, Allen and Hamilton Inc.

30. Wind, J. and Mahajan, V. (1997) "Issues and opportunities in new product development: an introduction to the special issue," *Journal of Marketing Research*, **34**, 1–12.

31. Griffin, A. (1997) "PDMA research on new product development practices: updating trends and benchmarking best practices," *Journal of Product Innovation Management*, **14**, 429–512.

32. Pye, D. (1978) "The Nature and Aesthetics of Design," *Marketing Science*, **9**, 3–15.

33. Foster, R. (1986) *Innovation: The Attacker's Advantage*. New York: Summit Books.

34. Anderson, P. and Tushman, M. L. (1991) "Managing through cycles of technological change," *Research Technology Management*, **34** (3), 26–34.

35. Ostlund, L. E. (1974) "Perceived innovation attributes as predictors of innovativeness," *Journal of Consumer Research*, **1** (September), 23–29.

36. LaBay, G. D. and Kinnear, T. C. (1981) "Exploring the consumer decision process in the adoption of solar energy systems," *Journal of Consumer Research*, **8** (December), 271–225.

37. Goldenberg, J., Lehmann, R. D. and Mazursky, D. (2001) "The idea itself and the circumstances of its emergence as predictors of new product success," *Management Science*, 47(1), 69–84.

38. Freeman, C. and Golden, B. (1997) *Why Didn't I Think of That?* New York: John Wiley and Sons.

39. Adler, B. and Houghton, J. (1997) *America's Stupidest Business Decisions*. New York: William Morrow and Company Inc.

40. Jorgensen, J. (ed.) (1994) *Encyclopedia of Consumer Brands*. Detroit: St James Press.

41. Goldenberg, J., Mazursky, D. and Solomon, S. (1999) "Creativity Templates: towards identifying the fundamental schemes of quality advertisements," *Marketing Science*, 18, 333–351.

42. Goldenberg, J., Mazursky, D. and Solomon, S. (1999) "Templates of original innovation: projecting original incremental innovations from intrinsic information," *Technological Forecasting and Social Change*, 6 (11), 1–12.

43. Goldenberg, J., Mazursky, D. and Solomon, S. (1999) "Creative sparks," *Science*, 285 (5433), 1495–1496.

44. Goldenberg, J., Mazursky, D. and Solomon, S. (1999) "Toward identifying the inventive templates of new products: a channeled ideation approach," *Journal of Marketing Research*, 36 (May), 200–210.

45. Kelso J. A. S. (1995) *Dynamic Patterns: The Self-Organization of Brain and Behavior*. Cambridge MA: MIT Press.

46. Ulrich, K. T. (1988) "Computation and pre-parametric design," MIT Artificial Intelligence Laboratory, Report No. AI-TR 1043.

47. Maimon, O. and Horowitz, R. (1999) "Sufficient condition for inventive ideas in engineering," *IEEE Transactions, Man and Cybernetics*, 29 (3), 349–361.

48. Nam P. Suh (1990) *The Principles of Design*. New York: Oxford University Press.

49. Stoll, H. W. (1986) "Design for manufacture: an overview," *Applied Mechanics Review*, 39(9), 1356–1364.

50. Holland, J., Holyoak, K. J., Nisbett, R. E. and Thagard, P. R. (1989) *Induction*. Cambridge, MA: MIT Press.

51. Hofstadter, D. R. (1985) *Metamagical Themas*. New York: Penguin Books.

52. Dowling, G. R. and Staelin, R. (1994) "A model of perceived risk and intended risk-handling activity," *Journal of Consumer Research*, 21, 119–134.

53. Morreau, C. P., Lehmann, D. R. and Markman, A. B. (2001) "Entrenched knowledge structures and resistence to 'really' new products," *Journal of Marketing Research*, 38, 14–30.

54. Calantone, J. R., Schmidt. J. B. and Song, M. X. (1996) "Controllable factors of new product success: a cross-national comparison," *Marketing Science*, 15(4), 341–358.

Index

Numbers in italics indicate *tables* or *figures*

advertising
  assessment of 151–2, 163–4
  common patterns notion 148, 151
  and Creativity Templates 148, 150–2, 164–6
  divergent thinking for 148
  examples
    Bally shoes 122
    British Lipton Tennis Championship 148, *149*
    Canadian Open Tennis Championship 148, *149*
    French Open Tennis Championship 148, *149*, 150
    Nike-Air® *120*, 151
  Janusian approach 152
  product space concept 151
  Raskin's psycholinguistic theory 152
  resonance approach 152
  symbols set 151
  Template approach 147–66
    Competition Template (Attribute in Competition, Worth in Competition and Uncommon Use) *156–7*, 161
    Consequences Template (Extreme and Inverted Consequences) *156–7*, 161, *162*
    Dimensionality Alteration Template (New Parameter Connection, Multiplication, Division, Time Leap) *158–9*, 163–4
    Extreme Situation Template (Absurd Alternative, Extreme Attribute and Extreme Worth) *154–5*, 161
    Interactive Experiment Template *158–9*, 161–2, *163*
    Pictorial Analogy Template (Replacement and Extreme Analogy) *154–5*, 160
    Template distribution in advertisements 163–4
aggregate diffusion parameters 17
Alexandria lighthouse 7–8, 59–60
Altschuller's creativity methods 33–4
amphibious car, source classification *214*
antenna in the snow 59–62
Attribute Dependency Template 59–75

accidents as sources of ideas 73
Attribute vs. Replacement Dependency 122–3
basic principle 63–5
between attributes vs. within attributes 74–5
cases/examples
  antenna in the snow 59–62
  baby-food 74
  candles, making a better 71–3
  chair 64–5
  chameleon 63
  common cold pain relievers 74
  diapers, disposable 74
  Domino's Pizza 66–71
  innovative lipstick 62–3
  Lighthouse of Alexandria 7–8, 59–60
  Post-It Notes® 134
components 64–5
cycles of dependencies 75
definition 9
Forecasting Matrix 88
generalization of 62–3
graphical depiction 64
implementation range 64
and macro operators 173
in nature 63
searching for 76
study of effectiveness for idea generation 188–92, 192–4
variables
  and components 65
  price as a dependent variable 68
  time as a relevant variable 65, 68
attribute space 168
awareness creation *see* market awareness

baby ointments
  analysis of, and the Forecasting Matrix 89–94
  Template effectiveness study 188–92
baby-food 74
Bally shoe ads 122
bandage
  invention of 201
  source classification *214*

bank, Template effectiveness study 185–7
blockbuster products, can Templates explain
        emergence? 180–4
    Post-It Notes® 182, *183*
    Sony Walkman® 182, *183*
brainstorming 45–53
    adverse effects 46–50
        deferred judgment creating a chaotic world
            48
        distractions 48
        fear of assessment 48–9
        free riding 48
        illusion of group effectiveness 49–50
        production blocking 47–8
    application 45, 46–7
    Conceptual Brainstorming 46
    EBS (electronic brainstorming) 51–2
    effectiveness, suggestions for 53
    focused brainstorming 52–3
    and nominal group methods 46–7
    Osborne's conceptualized approach guidelines
        45–6
    popularity, reasons for
        client impressing 50
        common organizational memory support 50
        competition over status 50
        diversification of ability 50
    Screening 46
    teamworking as an alternative 50–1
British Lipton Tennis Championship 148, *149*
butter patties 116–18

cake mix for home baking 124
Canadian Open Tennis Championship 148, *149*
candle, making a better 71–3
    use of Attribute Dependency Template 72–3
    new candles' benefits 73
    new candles' operation 72
    normal candles' operation 71
car phone speakers 3, *4*
car doors 101
Cellular Automata model 18
chair
    legless 126, 175
    as marketing example 24–5
    Replacement Template studies/operation
        109–14
chance and accidents, as sources of ideas 73
chewing gum (Orbit®) 142
cigarettes, smokeless 130
classification examples 214–16
close world condition 103–4
coffee drip filter, source classification *215*
common cold pain relievers 74
common pattern notion 151
Compaq Computer Corporation, keyboard stroke
        battery charging 2
Competition Template (Attribute in Competition,
        Worth in Competition and Uncommon
        Use) *156–7*, 161

Component Control Template 134–43
    and blockbuster products *183*
    definition/description 134–7
    examples
        computer screen with radiation filtering 135,
            136
        hair shampoo 135–7, 138–40
        Orbit® chewing gum 142
        Post-It Notes® 134
        women's stocking, radiation filtering by
            140–1
    and external components 138
    external effects 142
    internal effects 142
    limitations 142–3
    as a macro operator 175–6
    operational prescription 143
    thought process inherent in application 137–40
component space 168
components
    exclusion, when appropriate? 119
    internal and external 107–8, 138
        examples *108*
    intrinsic 117–18
    unsaturated intrinsic 117
    see also Component Control Template
computer screen with radiation filtering 135, 136
Conceptual Brainstorming 46
conditioner, source classification *215*
configurations of products *see* product
        configurations
Consequences Template: Extreme and Inverted
        Consequences *156–7*, 161, *162*
consumer awareness 17
consumer interaction 17
covariates, effect with success/failure prediction
        210–11
creativity
    Altschuller's methods 33–4
    creative people 32–3
    creative process 33
    creative spark ignition 30–1
    creative thinking and ideas 30, 33–5
    creativity code 30
    definition 29–30
    and genius and insanity 32
    lateral thinking 53
    mind mapping 54
    random stimulation 54
    'Six Thinking Hat' method 54
    see also brainstorming
Creativity Templates
    and Altschuller's methods 34
    application areas 1–10
    effectiveness studies
        Study 1: credit company 185–7
        Study 2: bank involved in savings accounts
            187–8
        Study 3: lateral thinking idea generation
            188–92

Study 4: drinking glass idea generation 192–4
Function Follows Form (FFF) principle 41, 83
and ideation 40
identification for advertising 150–2
for knowledge base manipulation 23
as macro operators 173–6
and Morphological Analysis 40–1
new candidate product prediction 41
restricted scope principle 41
validation of Templates theory 180–95
see also advertising; Attribute Dependency Template; Component Control Template; Displacement Template; Division Template; new product success/failure prediction; Replacement Template
credit company, Template effectiveness study 185–7
curtain replacement in Russian theater 101, *102*
customer feedback marketing considerations 15
customer reaction prediction 35–6
cycles of dependencies 75

Darwin's theory of evolution, comparisons 23
De Bono, E.
lateral thinking 53
'Six Thinking Hat' creativity-enhancement method 54
diapers, disposable 74
diffusion parameters, aggregate and individual-based 17
Dimensionality Alteration Template (New Parameter Connection, Multiplication, Division, Time Leap) *158–9*, 162–3
Displacement Template 124–33
advantages 130
and blockbuster products *183*
definition/description 124–5
examples
cake mix for home baking 124
legless chair 126
Mango® (Motorola's cellular phone) 125
sales catalogues of reduced size 131
smokeless cigarettes 130
soap that floats (Procter & Gamble) 131–2
Sony Walkman® 127
TV set with no picture 128–9
vacuum package 124–5
Waxless candle® (Ronson Company) 130–1
Function Follows Form (FFF) 126
implementation 128–9
as a macro operator 175, *176*
operational prescription 132–3
quantitative attribute, displacement of 131–2
Replacement Template, differences between 124, 125–6
and unbundling 127–8, 132
Division Template 176
and blockbuster products *183*

Domino's 30-min pizza delivery 6–7, 66–71
Attribute Dependency Template features
price (dependent variable) 68–9
table of dependencies *68*
temperature (independent variable) 68–9
time (independent variable) 68
competitors' strategies 67–8
market background 66–7
problems and prospects 69–70
Replacement Template implementation 104–5
doors in cars 101
dreamers 38
drinking glass, and the Forecasting Matrix 79–89

early determinants, why they can predict success 211–13
EBS, electronic brainstorming 51–2
Edison, T. A.
on inventions from hard work 37
inventive ability 2
swimming pool pump 1
effectiveness *see* Templates, effectiveness studies
Extreme Situation Template (Absurd Alternative, Extreme Attribute and Extreme Worth) *154–5*, 161
extrinsic sources for ideas 13–14

failure prediction *see* new product success/failure prediction
flat tire wheel change 102–3
following a trend 201
Forecasting matrix 76–97
Attribute Dependency Template with 88, 92
baby ointment analysis 89–94
active ingredient concentration 94
matrix for 90–4
odor and amount of excretion 92–3
product description 89–90
table of elements *91*
time and viscosity 93–4
concept 78–80
degenerated matrix 94–5
definition 94
drinking glass examination 79–89
definition 94
as a thinking exercise 85
heuristics for improving scanning efficiency 96–7
ideation process issue management 94–6
marketing state description 89
matrix for *80*
matrix elements 79
operational prescription 98
saturated matrix 95
table of elements *80*
variables
classification of 76–8
external 86–9
internal 81–5
freedom of thought, and ideation 39–40

French Open Tennis Championship 148, *149*, 150
Function Follows Form (FFF) principle 41, 83,
        106, 126

genius
    by Edison, T. A. 2
    creativity and insanity? 32
glass, drinking, and the Forecasting Matrix 79–89

hair shampoo 135–7, 138–40
hermit crab 104
heuristics, for improving matrix scanning
        efficiency 96–7

ideas, creative 30
    original, need for 33
    see also creativity
ideation
    and Creativity Templates 40
    and dreamers 38
    and freedom of thought 39–40
    as an ideas source 13–14
    ideation process issue management 94–6
    restricted and unrestricted considerations
        36–40
    and the Sparseness Theory 35
    study of Template approach as trainable 184–5
    synergetic effect 38
    and Templates theory validation 194–5
    Von Oech's tools 37
    see also creativity
individual-based diffusion parameters 17
innovation
    innovative lipstick 62–3
    Replacement Template for 105–9
Interactive Experiment Template *158–9*, 161–2,
        163
inventions, simultaneous 21

Janusian approach to creative advertising 152

lateral thinking 53
laws of product evolution 179
lead users 16–17
leadership status and innovative firms 13
lighthouse, Alexandria 7–8, 59–60
link (between components) 108
lipstick, innovative 62–3
liquid soap, mapping research on 179–80, *181*,
        182

macro operators 173–6
Mango® (Motorola's cellular phone) 125
manic-depressive illness and creativity 32
mapping research 179–80
    on soap-related products 179–80
market awareness
    diffusion of new idea awareness 19–20
    emerging need considerations 18, 19–23

manufacturers's responses 20
    S-shaped curve analysis 21–3
market needs and desires
    chair example 24–5
    and Darwin's theory of evolution 23
    shoe example 25–6
market research
    methods/techniques 15–16
    for new product design and success forecasting
        26
    for new products 201
    value of 26
market-based information 13
    macro level evaluation 17–21
    mapping/encoding 23, *24*
    micro level evaluation 16–17
marketing, creative and conventional approaches
        87
matrix *see* Forecasting matrix
mattresses, Template effectiveness study 188–91
mental invention 200–1
mind mapping 54
Morphological Analysis, and Creativity Templates
        40–1, 152

need spotting 200, 201
new product promotion companies 20–1
new product success/failure prediction 197–214
    early determinants, why success can be
        predicted 211–13
    environmental influences 198
    market acceptance and rejection 198–9
    project-level determinants 202–4
        changes in technology 203
        newness to the firm 203
        newness to the market 202
        product primary advantages 203–4
    research approaches 198
    Template prediction success studies 205–10
        effect of covariates 210
        patented products 205–6
        unified model of idea evaluation 206–10
    Template predictive power hypotheses
        early analysis performances 204
        idea source 204
        Templates 204
    Template-based approach 199–214
        early determinants 200–2
        following a trend 201
        idea sources 200–2
        market research for new products 201
        mental invention 200–1
        need spotting 200, 201
        solution spotting 200, 201
Nike-Air® shoes 119–21
    advertisement 151

operators *see* product operators
Orbit® chewing gum 142

oriented graphs 170
Osborne's conceptualized approach to
        brainstorming 45–6

patents
    product success prediction with Templates
        study 205–6
    radio-cassette locking 20–1
    and simultaneous inventions 21
Pepsi Clear®
    following a trend product adoption 201
    source classification 215
Pictorial Analogy Template: Replacement and
        Extreme Analogy 154–5, 160
pizza delivery see Domino's 30-min pizza delivery
Polavision®, source classification 215
Polo Harlequin® car (Volkswagen) 4–5, 6
Post-It Notes® (3M) 14, 134, 182
Pressman® (Sony) 14, 127
product configurations 108–9, 168–71
    characteristics: internal and external 168–9
    links 168–9
    space concept 151, 168
product operators 171–3
    inclusion and exclusion 171, 172
    linking and unlinking 171, 172
    splitting and joining 171, 172, 173
product-based information 13
products
    structural property similarities 35–6
    surface property similarities 35–6
propagation of awareness 17, 18
Ptolemy II, King 7

radio-cassette locking in cars 20–1
random events as sources of ideas 73
random stimulation, for creativity enhancement
        54
Raskin's psycholinguistic theory 152
Replacement Template 99–123
    and blockbuster products 183
    cases/examples
        antenna in snow 61
        Bally shoe ads 122
        butter patties 116–18
        candle, making a better 72
        car doors 101
        chair 109–14
        Domino's pizzas 104–5
        flat tire wheel change 102–3
        hermit crab 104
        Nike-Air® shoes 119–21
        Russian theater curtain replacement 101,
            102
        scanner 114–16
        Search for Extraterrestrial Intelligence (SETI)
            100
    close world condition 103–4
    components, internal and external 107–8
    configuration of product 108–9
    definition/description 4, 99
    exclusion, when appropriate? 119
    forbidden replacement 114
    Function Follows Form (FFF) principle 106
    implementation 104–9
    include suitable external component operator
        175
    for innovation 105–9
    join operator 175
    link (between components) 108
    and macro operators 174–5
    operational prescription 123
    Replacement vs. Attribute Dependency 122–3
    split and exclude operators 174–5
    step-by-step presentation 116–18
resonance approach to creative advertising 152
Russian theater curtain replacement 101, 102

S-shaped curve analysis 21–3
sales catalogues of reduced size 131
Sastratus of Cnidus, Alexandria lighthouse
        architect 8
scanner, and Replacement Template 114–16
Screening in brainstorming 46
Search for Extraterrestrial Intelligence (SETI) 100
shampoo, mapping research on 179–80, 181, 182
shoes, as marketing example 25–6
'Six Thinking Hat' creativity enhancement method
        54
smokeless cigarettes 130
    source classification 215
snacking doll, source classification 216
soap-related products, mapping research on
        179–80, 181, 182
soap that floats (Procter & Gamble) 131–2
solution spotting 200, 201
Sony Walkman® 14, 127, 182–3
source classification examples 214–16
space
    attribute space 168
    component space 168
Sparseness Theory 35
split and exclude operators 174–5
spontaneous discovery awareness creation 18
structural properties, similarities between
        products 35–6
success prediction see new product success/failure
        prediction
surface properties, similarities between products
        35–6
surprise and regularity, balance between 35–6
symbols set 151
synergy, and ideation 38

teamworking, as an alternative to brainstorming
        50–1
Templates
    detection and utilization in products 35–6

Templates (*cont.*)
    effectiveness studies
        Study 1: credit company 185–7
        Study 2: bank involved in savings accounts
            187–8
        Study 3: lateral thinking idea generation
            188–92
        Study 4: drinking glass idea generation
            192–4
    explaining/predicting blockbuster product
        emergence 180–4
    trainable as an ideation method? 184–5
    see also new product success/failure prediction
Thermos® flask, source classification *215*
trend identification 17
TV set with no picture 128–9

ultrasonic device against cats, source classification
    215
unbundling, and Displacement Template 127–8,
    132
unified model of idea evaluation, success
        prediction with Templates study 205–6
unpredictictability 35
unwaxed candle, source classification *216*

vacuum package 124–5
variables
    classification of 76–8
    dependent and independent 78
    external
        defined 77
        examples 77, 86–9
    internal
        defined 76
        examples *77*, 81–5
    and the Replacement Template 108–9
Vaseline®
    invention of 201
    source classification *215*
voice of the customer 15
Von Oech's ideation tools 37

Walkman® (Sony) 14, 127, 182–3
Waxless candle® (Ronson Company) 130–1
wheel change for flat tire 102–3
Wirefree Ltd, car cellular phone speakers 3, *4*
women's stocking, radiation filtering by 140–1
word-of-mouth awareness creation 18
workshops of effectiveness *see* Templates,
        effectiveness study workshops